P9-CKQ-957

THE DARK NIGHT SKY

THE DARK NIGHT SKY

A Personal Adventure in Cosmology

DONALD D. CLAYTON

A Demeter Press Book

QUADRANGLE / THE NEW YORK TIMES BOOK CO.

The author is grateful to the following for permission to quote from—
T.S. Eliot. *Four Quartets: East Coker* (Harcourt Brace Jovanavich, Inc., New York, N.Y.)
Albert Einstein, *Ideas and Opinions* (Crown Publishers, Inc., New York, N.Y.)
Thanks are also due to those who have provided photographs, as indicated in the captions. Uncredited photographs are by the author.

For information, address: Quadrangle/The New York Times Book Co., 10 East 53 Street, New York, New York 10022. Manufactured in the United States of America. Published simultaneously in Canada by Fitzhenry & Whiteside, Ltd., Toronto.

Library of Congress Cataloging in Publication Data

Clayton, Donald D
The dark night sky.

1. Cosmology. I. Title.
OB981.C62 523.1 75-9213
ISBN 0-8129-0585-7

for Annette

CONTENTS

FOREWORD

It was the custom some years ago at the California Institute of Technology for annual challenges to a game of baseball to come from one laboratory to another. My friend William A. Fowler once told me that no games were ever lost by this particular laboratory once a graduate student named Donald D. Clayton appeared on his roster of players. Don might, with his physical capabilities, have followed a career very different from his present one. He might have followed his father, for example, and become an airline pilot. But Clayton became a scientist. Why? The reader will get some feeling for the answer to this elusive question from the outset of this book.

Scientists come from families in all walks of life. They are the children of airline pilots, farmers, construction workers, businessmen, store-keepers, or wool merchants. . . . There is just no limit to the variety. Almost the last thing they ever tell each other are details of their personal family background because such details are usually irrelevant to the matters in hand. While science spans all other human activities, it stands apart from them. This isolation creates a peculiar difficulty for the scientist, should he turn for a while from science back to the society of his early years, or to society in the large. How shall he explain himself? How shall he describe his achievements and those of his fellow scientists? These questions turn out to be so awkward that many scientists simply do not attempt to answer them. Yet here in this book Don Clayton has done so, particularly from the point of view of motivation, which is something different from the usual expositions of science. In effect, Clayton has addressed himself to the problem of what it is that drives the scientist helter-skelter along a road which nobody except himself and perhaps a few colleagues can perceive at all. He has done so partly from his own experience and partly from historical examples.

I have always felt the lucky people in life are those who know with complete certainty exactly what they wish to do. Happiness is not for those who are uncertain and unsure, not for ditherers. Scientists figure prominantly among those who know just what they wish to do. By and large, they are lucky people. Don Clayton seeks to tell you why this should be so.

Fred Hoyle
December 30, 1974

PREFACE

In what follows I share parts of my life with the reader. The intellectual adventure that I want to share lies at the root of human thought. It is the adventure of man's comprehension of his universe. The associated science is called cosmology, and it is my purpose to create an appreciation of its main facts. The observational basis of our knowledge of the universe comes from the science of astronomy —not just optical telescopes looking at stars and galaxies, but every conceivable new technique for receiving and recording information from "out there." Cosmology is more than astronomy, however, for it is also the mind of man struggling with the logical and philosophical meaning of existence. It is the attempt to piece together the whole history of existence from the snapshot of our world today. I myself am a scientist working daily on these problems, so the most I can promise the reader is that the contemporary scientific ideas concerning man's view of the universe will be rightly told. Any deeper relevance will be up to the reader to grasp himself. The telling of it involves my life because that is the way it happened to me, and if all goes well between us, sharing my experience will be a way of partly sharing the human experience. I do not pretend to cover all of the important ideas in cosmology; that is the task of a textbook. Nor do I pretend to give opinions free from my personal bias, for to do so would defeat my purpose. I am content to say the things I have to say about those ideas that never leave my mind.

The desire to write a book for people is due to people. Fred Hoyle appears throughout this book because his unswerving dedication to the search for meaning has been an inspiration to me. The vast expanses of Texas of which I am so fond are also part of my story, for they were and are my most formative contacts with earth and sky. Annette, my wife, is there, too, because we shared it all with love.

THE DARK NIGHT SKY

CHAPTER I
THE DARK NIGHT SKY

The dark night sky is the messenger of a great adventure of mankind. It is an adventure of the mind and of meaning. Its structure began focusing for me personally not so many years ago on an April evening. It grew within me somewhat like my parents' maple tree leafs out in spring—first a little on that branch, then some more at the top, then here, then there, until finally the branches blend together in an explosion of green. The adventure became the focus of my intellectual life, and the telling of it prompts all that is to follow. To begin sharing it with me, you must join me late that April night driving down U.S. Highway 84 as fast as my old green Pontiac could safely go. I had tanked up in Lubbock and was now about 15 miles south of Post on my way toward Sweetwater. My spirits were quite high as I anticipated the upcoming holidays at home, half a continent from the rigors of graduate school.

I was then a research student at the California Institute of Technology, pursuing a course of study and research that I hoped would lead to a Ph.D. in physics. During the afternoon I had often thought about what progress I was making toward that goal, and I admitted that I was not doing as well as I had hoped. It was not that I hadn't the ability. My objectives seemed weakly focused on nuclear physics and lacked the motives to justify the sacrifices and hard work that would bring success. Instead, I kept thinking of the things I was learning from astronomers and of my curiosity about what was happening in the centers of stars. So fascinated was I by the latter question that I had all but decided to shift my research efforts from pure nuclear physics to nuclear astrophysics, which I would describe as the application of knowledge of nuclear physics to problems of astronomy and cosmology. Just now, however, my eyes fell fondly on the expanse of Texas. Only an occasional ranch house light, an occasional gas flame atop an oil well, and the headlights of an occasional car altered the darkness of the night. It was quite a change from the traffic of La Cienega Boulevard, or even from Pasadena, and I liked the change. I felt an exultation at the reawakening of my roots in my home state. Except for the fences surrounding it, this west Texas land is almost a wilderness. I am always moved by it.

You have to have seen the rabbits to believe them. I always wondered why the rabbits like the roadside verges so well, or if there were as many rabbits throughout that vast dark land as I now saw lining the highway. Nibbling away in groups of three or four or more, their eyes flashed like gems in my approaching headlights. They would watch attentively until I was fifty yards away, when they would bound off, ears down, on their jagged paths to safety. Most headed for the fences, but many, confused, chose paths that crossed the road. As I swerved in repeated attempts to dodge them, my mind recalled my uncle in his pickup truck as he always tried to hit them, saying: "There, by God, that's one less rabbit to put up with!" I was always sickened by the thud. Even trying to miss them that night I hit two, and the many other rabbit corpses I saw were testimony that many drivers did not share my concern for rabbits.

I don't know why I suddenly sought to see if I could approach them better with my headlights off. I punched them off and realized that I could still see the long straight deserted road in front of me. Despite the absence of the moon, the road noticeably reflected what little light there was. Thus reassured I slowed down and drove along through the dark, now straining much harder to see the rabbits and their reactions. The experiment itself was not noteworthy; subjectively I decided that they were more in fear of the engine's roar than of the headlights. I had slowed only to observe one large group as I eased by, and then I looked up through my open car window.

It is no exaggeration to say that my life has not been the same since that night. As I tell it now, fifteen years later, it remains vivid in my memory. What I saw outside my car as I stopped and climbed out was the clear blue-black coat of heaven pierced by its own burning lights. Who among us does not know that view? I had often looked at the stars during my childhood, and now I was even studying their theoretical structure in one of the greatest of universities, but somehow I was unprepared on this occasion for the impact of the view which spread itself above me. The moonless night was very black. Never had I seen so many stars. The longer I looked, the more I saw as my eyes dilated to accommodate that darkness. Where first I noticed none I soon saw whole clusters of faint and twinkling objects. Above the western horizon Orion and Canis Major shone like a display, and the Hyades and Pleiades stood bunched in close array. Sirius was overwhelming, as was Procyon in Canis Minor. In Orion bright Rigel was close to setting in the west, and how majestically it shone through that dark clear horizon. No artificial lights disturbed that view. The Big Dipper was fixed straight above, and Draco's tail wound menacingly around Ursa Minor. The sparser eastern sky was dominated by Arcturus and Vega. And toward the south the vast area occupied by Hydra and its satellite asterisms seemed bleak and dark in comparison with the brilliant stars of winter now leaving in the west. Occasionally the stars twinkled so that hey seemed to dance and blink as their rays penetrated the clear air turbulence of the earth's atmosphere. Some, like Rigel, were very blue whereas others, like Betelgeuse, were noticeably red.

As I continued looking, two thoughts thrust themselves upon me—one annoying and one puzzling. At my annoyance that I had not for years witnessed such a view, I realized the obvious. In Los Angeles you frequently cannot see a thing! Even if there were no smog, the blazing city lights lighted the sky and obscured its true visible contrasts. I recalled astronomers at Mt. Wilson Observatory telling me that the atomic lines of the element mercury were prominent in every spectrum of every object. Everyone laughs. It comes from the sky reflection of the light from the mercury in the gas discharge tubes that give Los Angeles its eerie nighttime glow. I was to hear fifteen years later while visiting the construction of the magnificent 150-inch telescope at the Kitt Peak National Observatory high in the Papago Indian Reservation of southern Arizona that the lights of mushrooming Tucson were already a similar threat. I feared then and fear more now that it will be harder and harder to get a really good outward look—especially for urban man. Perhaps that is no significant loss. Growth is commonly regarded as being our most important product.

The puzzling thing was that in spite of the serenity of that sky, my senses seemed tense. I had noticed it before in spending nights out in the wilderness. A sense of danger seems to exist. Instinctively I alerted myself to be sure no savage creature could sneak up on me. In retrospect, I was giving those rabbits a lot of credit, but the fear sprang from some deep unconscious level. It did not escape me that the evolution of man consumed several billions of nights of stark terror. A mere few thousand years away from nature in the raw have not eliminated that sharp instinct—at least not in me. I could not but imagine that in those billions of nights of the evolution of my own ancestors, the only constant nightly sentinels were those I now saw above. During all those days on animal-filled plains we were busy killing and digging for food, or so I am told, but at night alone with our thoughts we would have seen only that sky. How many sleepless nights were spent huddled in some hidden spot, constantly alert to danger, and constantly wondering at that dark night sky and awaiting the signs of dawn. I think now of Robert Ardrey's fascinating *African Genesis,* and I also see Stanley Kubrick's film *2001* with those terrified apemen huddled together in some slight cave —looking and waiting. Is it any wonder that man's earliest mysticism, earliest religion, and earliest science all centered on the heavens? That nightly view has dominated our psychic growth. Are we, dwellers of modern lighted cities, being cut off from that evolutionary heritage? I felt all at once as if I understood the almost fanatic curiosity of my astronomer friends. What could be more natural? Night after night they sit in dark cages located on huge modern telescopes looking outward for understanding. I must admit that then and there I made an emotional contact with my scientific life. It felt good.

I cannot pass on without remarking that this decisive experience for my own career has a parallel in the poorly understood thought processes by which scientific leaps are made. One observes a perfectly commonplace thing—in this case the night sky—and sees in it

The main portion of the constellation of Orion, dominated by bright blue Rigel at the lower right and red Betelgeuse at the upper left, as blown up from an ordinary photograph taken with a telephoto lens. The famous Orion Nebula, where new stars are now being born, is in the close group of three starlike objects just below the three bright stars of the middle belt.

a whole new connection that was not "ripe" in the many earlier occurrences. Ordinary facts, like the summer sun rising to the north, like the darkness of the night sky or the blueness of the day sky, like the continued existence of radioactive uranium, like the equal rates of fall of a stone and an iron ball, and many others—these acquire special new meaning when a mysterious connection is made within the brain to a deeper set of meanings. Such events are precious and unpredictable, and are probably related to many forms of intellectual creativity. In this case, like Saul's on the road to Damascus, it was the motives of my own life that were changed in a flash.

Another disquieting experience had happened in October of the same school year. During the weekly Thursday physics research colloquium at Caltech, it was remarked excitedly that Sputnik 1 would pass overhead visibly during sunset. As a body, 200 physicists—eminent professors, post-doctoral assistants, and students—filed to the roof of Bridge Laboratory to watch for it. Many sudden shouts and pointing fingers later, it came like speeding Venus across the twilight. The chatter died quickly away, and in five minutes of silence we watched its pass. Just as suddenly, it was gone, and we all stared dumbly at each other, just barely comprehending that mankind would never be the same again.

As my thoughts returned on that April evening to the sky itself, a new aspect of the light caught my attention. Between Canis Minor, marked by Procyon, and Canis Major, marked by Sirius, spread a band of light. My eyes followed it all the way from the southern horizon up to a point somewhat west of overhead and on toward the northern horizon. That continuous glowing band of light occupied no more than about one tenth of the visible sky overhead, but the longer I looked the more commanding of attention it became. It was as if a giant painter had slapped a single stroke of white paint from pole to pole. Many stars were visible in the band because they are relatively nearby, whereas the band itself has such a distant white glow that it is milky in appearance. I have no idea who first called it the "Milky Way," but I could see that the name has stuck because it is so apt. Nor could I then see that considerations of the Milky Way were to occupy me till today. I cannot but wonder at this last coincidence because my momentary fascination with that band played no role in my conscious thoughts as my subsequent scientific interest developed from a quite theoretical perspective.

Star clouds in the Milky Way in the direction toward the center of our galaxy. This time exposure with a 250 mm telephoto lens shows many more stars than the naked eye can see. (Hale Observatories)

Although the ancients knew the Milky Way, they had relatively little to say about it. I find that curious, considering my own fascination, but I suspect it was because it has so little structure evident to the unaided eye that priests and poets could not easily fabricate myths about it. It was so much more appealing to them to contemplate the groupings of bright nearby stars. Like some heavenly Rorschach test, their relative positions could suggest animals, stories, fears, heroes, and gods. The groups, or constellations, acquired names, and thereby personalities and mystic powers. The constellations are, in fact, deceptive because their prominent stars often lie at vastly different distances. Many, it turns out, actually are close together because they came into being together in the same dense cloud of interstellar gas—heavenly brothers from the same mother cloud. The ancients could not resist associating anthropocentric meanings to their positions, even though, unknown to them, many just appear to be close together.

When I was a boy I was told that there were more stars visible in the heavens than I could ever count. Just glancing at the sky that night one could believe it, although it's not true. Concentrating on a small imaginary box in the sky, I found I could count every visible star within that box. I estimate that the whole visible sky is no more than a few hundred such boxes. By careful counting the eye can see about 6,000 stars in the whole sky. As Galileo dramatically showed near the beginning of the seventeenth century, a telescope brings many more stars within view. With even a modest telescope, the same small box now shows thousands of stars instead of only tens. Larger telescopes show even more. The Milky Way itself is easily seen to be composed of multitudes of distant stars. About a million of them have been catalogued by astronomers in various ways. Routine telescope photographs of the sky show about 100 million uncatalogued individual stars. By inference we now know that the Milky Way has many more stars yet that are hidden by the obscuring gas that exists in great clouds along the plane of our galaxy—about 100 billion, or 100,000 million, stars in all. To count them at the rate of one per second, mankind would have to have begun counting about 1,000 years before the birth of Jesus of Nazareth and have continued till today. That's a very large number—seemingly infinite in terms of everyday experience—but actually not infinite. The mind calls for a rest, but the real shocker must be told. When the

60-inch telescope on Mt. Wilson was completed around 1913, it confirmed that faint patches of light among the stars are actually distant galaxies—other "Milky Ways" for us to see, other groupings of 100 billion stars. The number of these galaxies is so huge it seems they will never be counted. They keep appearing, dimmer and dimmer, as new telescopes look further and further. There's the basic question: *What is it all?*

I decided to spend the night nearby in Sweetwater, where I found a small motel on old U.S. Highway 80. I slept so well that I didn't get away until 8:30 the next morning. A brilliant morning sun shone in my eyes as I headed eastward. Listening to country music on my car radio, many miles rolled easily by before I noticed that the highway was clean. Only an occasional dead rabbit could be seen. The vultures too had been watching the night sky, and they had circled into action at dawn's first light.

CHAPTER II
SILENT WITNESSES

If I had been told as a child that the apparent motions of the sun and moon stimulated agricultural planning, science, and religion, I would have believed it. I had not yet been educated in the miracles of modern science and the associated tendency to regard prehistoric human beings as apes. Nor had I yet been insulated from the formative nightly view by the complex modern megalopolis with its lights and congestion. As a child I rose when the sun rose and I went to bed after it set. Sunrise and sunset were two of my favorite moments—somehow mystical moments, when my imagination, more often than not, took flight. My days were intimately linked to the sun, and I could have easily regarded it as a God, had I been so instructed. As a teenager I arose every morning before sunrise to deliver newspapers by bicycle, and I saw that magic moment daily for four years. I can't recall when I came to know it, but I knew that in summer the sun rose and set toward the north, and in winter it rose and set toward the south. In spring and fall it rose directly in the east. I didn't understand why this was so, but I was aware of it—I suppose, though I am not sure, because the north Dallas streets of my newspaper route ran due east-west. In my mind's eye today I associate hot weather (in one August alone we had 21 days above 100° F) with sunrise on the north side of those streets and cold weather (my hair, combed with water, often froze on my head) with the southerly sunrise.

This is of interest to me now because any contemplative person who regularly watches risings or settings of the sun might easily notice their periodic northerly and southerly excursions. My farming uncle knew it, and I infer that the knowledge is common among farmers. I myself was not aware that the moonrises and moonsets showed similar, though somewhat more complicated, migrations; but I never watched the moon much. I did not systematically observe the night sky because I was doing my homework and preparing to go to bed. I can, however, easily believe that prehistoric outdoor man, with less to distract him from nightly gazing, would have noticed it. What one has just noticed is likely to be told to others—but perhaps not *all* others, for the enlightened few might find some advantage in their mysterious secret!

The secret is exposed by placing a model of the earth, a globe, in the beam of a spotlight which can be regarded as a model sun. As the globe rotates once a day about its own polar axis, each point on its surface rotates easterly so that, as it moves from shaded night into illuminated day, the spotlight first hits the point from the eastern horizon. The North Polar axis of the earth is not exactly perpendicu-

Sunrise at Stonehenge in midsummer with a full moon

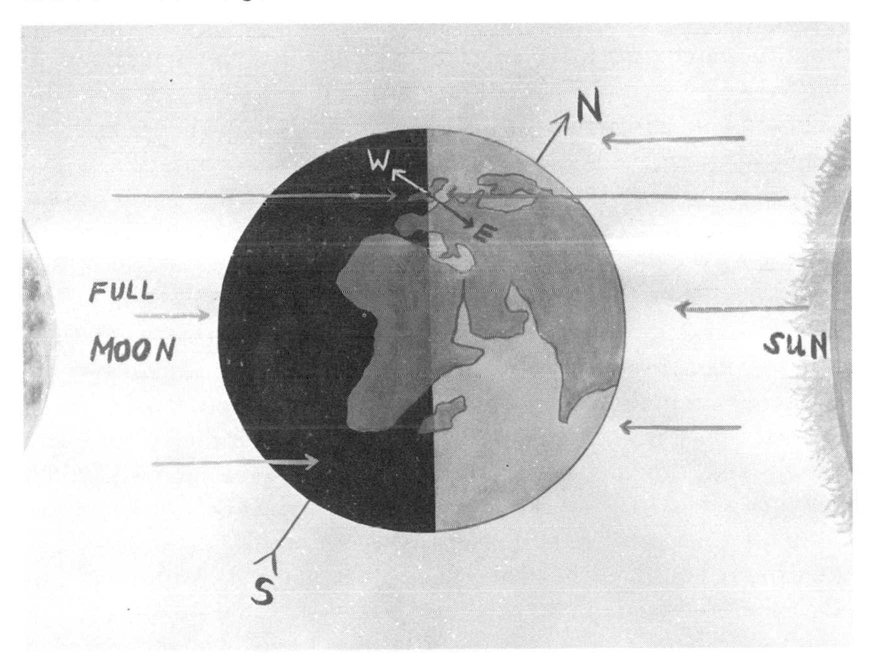

lar to the plane of the earth's yearly orbit around the sun, however, so the spotlight should not generally be directly overhead on the equator. At midsummer (for the northern hemisphere) the earth's North Pole is tilted by 23° toward the sun, so the sun appears directly overhead at a point whose latitude is 23° north of the equator (the Tropic of Cancer). To someone even further north, say, in North America or Europe, the noontime sun cannot be directly overhead but must appear to be south of the line straight up from the earth's surface. Now imagine a summer sunrise in London, when that point has rotated out of darkness into the first beam of the spotlight. Because the North Pole is tilted in summer toward the spotlight, the first rays of dawn will graze the earth from a direction to the *north* of east. By noontime the sun line is south of directly overhead, but at sunset its last grazing rays will again reach London from a direction somewhat *north* of due west. Thus it is that sunrise and sunset in summer in London occur about 39° north of due east and due west respectively. Much more northerly yet, the summertime North Pole is illuminated by perpetual daylight as the earth rotates.

With its large mass rotating daily about its polar axis, the earth behaves like a gyroscope. Such a spinning mass possesses a physical quantity called "angular momentum," and by Newton's laws of motion we now understand that both the rotation axis and the magnitude of this angular momentum will remain fixed unless a torque is applied to the object to change it. No torque acts on the earth moving freely through space, so the angular momentum associated with its spin does not change direction as the earth progresses on its annual orbit of the sun. Therefore six months later and in midwinter, when the earth has moved to the diametrically opposite side of its orbit around the sun, the North Pole of the earth will then point 23° *away* from the solar direction. The midwinter North Pole itself is always in the shade of the earth, sitting in perpetual night as the earth spins daily. Because the North Pole axis is then tilted away from the sun, the first rays of morning light in London appear 39° to the south of east. Thus it is that sunrise and sunset occur toward the southeast and southwest in midwinter London. The southern hemisphere, on the other hand, is experiencing its "summer" because the South Pole is now tilted toward the sun.

I cannot claim that I understood all this as I delivered the morning newspaper; nor did my farming uncle understand it as he planned the seasons of his crops. But it is interesting that the facts could be apprehended without any theoretical understanding. It is also fascinating that one of the few things we know of prehistoric Stone Age man is that he also knew these heavenly motions. Not only did he know of them, but he built huge stone observatories (or temples) to measure them. The most fascinating of these to me is Stonehenge.

I first saw Stonehenge on a sudden impulse. My wife Annette and I had been on a touring holiday through the coast of Devon in western England, and we had begun the return trip toward Cambridge where I was a Visiting Fellow of Fred Hoyle's Institute of Theoretical Astronomy. Coincidentally, the first scientific account I had heard of Stonehenge was from Fred Hoyle when he gave a Friday night lecture on it in Caltech's Beckman Auditorium in the spring of 1967. I remembered that lecture while we were looking at the cathedral in Wells, and we then decided on a detour—slightly south of east toward Warminster, where we took the A344 toward Amesbury. The road leads over a beautiful rolling plain. The view gives the impression of being high, probably because the visible horizon is so much more distant than in the many nearby valleys. Two miles west of Amesbury we encountered Stonehenge—huge assorted slabs of stone standing upright on the grassy plain. According to recent discoveries it seems to be the remains of a neolithic astronomical observatory and ceremonial temple, where Stone Age man observed the positions of the risings and settings of moon and sun.

We reached the monument itself with the aid of much more modern stonework, the concrete tunnel leading under the A344 from the concrete parking lot on the opposite side of the road. At that time there were perhaps fifty people of widely differing types milling about among the stones. There is now a constant stream of visitors because the monument holds a strong grip on the public imagination. Her Majesty's Ministry of Public Buildings and Works estimates a quarter of a million annual visitors, which averages almost to 1,000 a day. No one is quite sure why they all come because very few persons (including myself on that first visit) have a very good idea what they are looking at. Indeed the major activities of the visitors would seem to be absent-minded gazing upward at, or gently kicking the bottom of, a huge slab of stone, photographing a friend in some incongruous relation to a stone, or simply wandering about. This is no criticism, for they share the puzzlement of all men since recorded history began. About 2,000 years before Christ, a band of Stone Age men miraculously assembled these enormous stones into an intricate structure capable of measuring the seasons and even of predicting eclipses of the sun and moon; but within a few hundred years, due to some conspiracy of events, they forgot how to use it. And thus it has stood until modern times, with assorted carelessnesses, vandalisms, and natural weathering to make it all the harder to comprehend.

Visually Stonehenge is dominated by two concentric circles of stones enclosing two series of stones arranged as horseshoes, all about a common center. The parallel horseshoes open toward the northeast, where 256 feet from the center stands a huge stone, now called "The Heelstone." Its large size, an estimated 35 tons in weight, is typical of the large stones called "Sarsens" that also form the outer stone circle and the outermost horseshoe. Because the hill slopes gently downward from the center of the circles to the Heelstone, its 16-foot-high tip appears in line with the distant horizon. That might seem an accident, except that on the morning of June 21, the longest day of the year (called the summer solstice, which is also the day when the sunrise occurs farthest to the north in its annual north-south cycle), the sun rises on the northeast horizon from the tip of the Heelstone!

Later I read Gerald S. Hawkins's book, *Stonehenge Decoded*. He showed that the other prominently suggestive lines of sight correspond to the extreme risings and settings of not only the sun but also the full moon. Preoccupation with the full moon is not surprising, for poet and legend alike bear witness to its magic powers. The great prominent upright stones did not yet exist in the earliest version of Stonehenge, which was completed about 1850 B.C. At that time there existed ten large sighting stones in addition to the Heelstone, so arranged as to suggest sixteen important lines of sight. Each of these sixteen lines of sight points toward one of the twelve known positions of extreme northerly or southerly risings or settings of the sun or full moon. The reason there are twelve extreme positions is that the full moon has eight such positions whereas the sun has only four, its northernmost summer sunrise and sunset and its southernmost winter sunrise and sunset. Full moon occurs when the sun shines on the entire surface facing the earth; that is, when the moon, in its 28-day orbit of the earth, lies approximately on the opposite side of the earth from the sun (not *exactly* on the opposite side or we would have a *lunar eclipse*, when the full moon falls in the shadow of the earth). Since the North Pole tilts toward the sun in midsummer, it necessarily tilts *away* from the full moon. Thus, although midsummer sunrise is in the north of east, midsummer full moonrise is in the south of east. The full moon would also have four extreme positions if the plane of the moon's orbit around the earth coincided with the plane of the earth's orbit around the sun; however, it is obvious that in that case the full moon would be eclipsed in every month. In fact the lunar orbit is tilted with respect to the earth's orbit with two major results: (1) most full moons are not eclipsed and (2) the southernmost position for midsummer full moonrise oscillates on an approximately 18⅔-year cycle between declinations of $-29°$ and $-19°$. It is these two extremes that double the number of extreme positions of the full moon. In addition three stones were placed whose sighting lines correspond to the solar equinox, which occurs on the day in spring and fall when the sun rises directly in the east and to the risings of the full moon at the equinoxes. These two dates fall halfway between midsummer and midwinter when the earth has revolved in its orbit to those points where the North Pole is perpendicular to the solar direction. These dates are of obvious importance in constructing an astronomical calendar. The day, the week, the month, and the year all owe their existence as concepts to those early people who counted the recurrences of the sun and moon. The sun gave us daily light and dark and annually winters and summers. The moon takes seven days to change from full moon or new sliver into a half illuminated globe and vice versa. It also completes the whole cycle in a lunar month. We glibly use these terms with no remembrance of their fundamental origins and the effects they must have had on the human psyche.

When the great 97-foot circle of thirty equally spaced Sarsen Stones capped with heavy stone lintels and the great inner Sarsen Horseshoe were added in a later wave of building at the advent of the Bronze Age, about 1650 B.C., these significant directions of the original Stonehenge were clearly appreciated. This Sarsen Horseshoe is constructed of five so-called trilithons—five pairs of huge vertical slabs of stone capped with five horizontal lintels. The verti-

The view toward the heelstone from the center of Stonehenge. One midsummer's day the sun rises from the tip of the heelstone. (S. G. Perrin)

cal stones weigh a staggering fifty tons each! The archways within each trilithon are surprisingly narrow for such big structures. Since the openings between the slabs average about 12 inches, whereas their thicknesses average about 2 feet, the view through them is necessarily constrained to reveal a limited number of archways in the large circle of Sarsens. The views through these double archways are directed toward the extreme risings and settings of the sun and moon! These can hardly be coincidences, and I for one feel slightly humble in their presence.

To further confound those who would believe these alignments occurred by chance, one should note the special latitude in southern England where Stonehenge was constructed. The four so-called station stones of the original Stonehenge form a large rectangle whose center is at the center of the circles. One side of the rectangle is parallel to the center-Heelstone line and thus points toward the rising sun in midsummer or the setting sun in midwinter. The perpendicular sides of the rectangle point toward the extreme rising of the full moon at the same time. The perpendicularity of these two directions is peculiar to Stonehenge. It lies at just that latitude where these extreme risings occur at right angles on the horizon. Had Stonehenge been constructed in northern England or on its southern coast, a rectangle would not have worked. Since comparable but less advanced neolithic stone observations exist elsewhere as far north as Callanish in the Hebrides and south just across the channel in France, it would seem unlikely that the special latitude chosen for Stonehenge could have been accidental.

One other major part of the original Stonehenge has caused much puzzlement and speculation. Before the importation of the spectacular stones for the circles and horseshoes, the much simpler view was contained within a large circular earthen bank, 320 feet in diameter. The chalk and rubble and dirt for this bank was excavated from a ditch just behind the bank. In an aerial photograph they still comprise the dominant visual feature of the monument. The mind's eye must visualize the awesome effect of this thick circular wall of white chalk then standing almost 6 feet above the ground. Within this bank exists an accurate circle of fifty-six holes filled with chalk. John Aubrey, the antiquary, recorded seeing earthen cavities above these spots in 1666. When later excavation revealed them they were named "Aubrey Holes."

Why did those people, whoever they were in England about nineteen centuries before Christ, dig these holes and then fill them up again with chalk rubble? Why fifty-six holes? Why were they evenly spaced in an accurate circle 284½ feet in diameter? Hawkins and Hoyle have both published arguments that the Aubrey Holes were utilized to predict the eclipses, which can occur only when the sun, moon, and earth lie almost on a straight line. This does not occur each lunar month because the moon's orbit is inclined, but the line where the plane of the moon's orbit makes the necessary intersection with the plane of the earth's orbit revolves in very nearly 18⅔ years. Curiously, the number of Aubrey Holes is the smallest whole number that is a whole multiple of this period; i.e., $56=3\times18\frac{2}{3}$. Thus, if a stone were moved three holes per year around the circle, it would revolve at the same rate as the line of lunar nodes where the moon crosses the ecliptic plane. Another coincidence? Yet such information is necessary to predict eclipses. The fifty-six chalk holes can even represent the revolving of *all three* heavenly objects and so give an accurate prediction of eclipses. Another stone moved two holes each day completes the orbit in 28 days, very nearly the period of the lunar orbit. A third stone moved two holes each 13 days completes the orbit in $28 \times 13 = 364$ days, very nearly the period of the earth's orbit around the sun. Hoyle has shown how the small inaccuracies are easily recalibrated periodically to keep the entire eclipse computer in perpetual synchronization. The motion of markers as models for the motions of heavenly bodies explains why pains were taken to place the Aubrey Holes in such an accurate circle and with such a great size that it connotes an astronomical distance. The great chalk wall behind it was the wall of heaven, within which we and our heavenly gods are enclosed. That is, Stonehenge was a working model of the universe! A cosmology.

It staggers the mind. Nineteen centuries before the Christian era Abraham in Mesopotamia was making his biblical covenant with God. And here in England unknown people were building working models of the gods. It stretches our credulity, but it is even less probable to imagine that a computer capable of all these things was constructed by accident. That would be like the monkey at the typewriter accidentally punching out *Hamlet*. Perhaps we are reluctant to admit that these prehistoric Stone Age people knew what very few of us know today. But we must remember that they were familiar with the night sky that we scarcely see. We are prone to doubt their ability to accomplish such a cultural effort because we have difficulty in visualizing its astronomical causes, which we find explained in erudite books along with DNA and the atom. It was, however, quite another matter for people nightly watching the moon rise beyond a distant mountain peak or daily seeing the sun rise along a narrow easterly valley. It seems that we must upgrade our opinion of the intellectual status of Stone Age man.

The construction of Stonehenge was a massive enterprise. It must have been extremely important to these prehistoric people because it was physically laborious and required a sustained level of intellectual planning and group effort for several centuries. It was not a matter of some shepherd piling up a few rocks in his field. The eighty or more so-called bluestones that form the inner circle and the inner horseshoe originated within a small area of the Prescelly Mountains in Wales! Thus, the overland and overwater route for these stones, which weigh up to 5 tons each, is about 240 miles. Teams of men probably pulled the stones on sledges over rollers made of tree trunks. Hawkins estimated that about sixteen men per ton of stone could move them about a mile per day by this method. The eighty huge Sarsen Stones comprising the outer circle and outer horseshoe probably came from the Marlborough Downs about 20 miles to the north. These stones average 30 tons in weight and their transport would have involved perhaps 10,000 human work-years. The shaping of the stones and their erection was another big effort. Hawkins calculated that, all in all, 1,500,000 human work-days of labor were involved in the construction of Stonehenge. The scientific planning and logistics must have demanded the continuing contribution of many generations of the land's most gifted individuals. In cost and meaning it was to them what the Apollo program has been to us, although it took longer to accomplish. They probably even called it something like the "Apollo Program" and regarded it as their special cultural exploration.

Why they took the trouble is not hard to understand. Life was hard and especially dependent on the benevolent gifts of the summer season. In analogy to the natural cycle of vegetation, farming profits greatly from a knowledge of the seasons and the best times to plant seed. Surely one use was a dependable calandar. And beyond this it seems to be a psychic drive of human beings to reduce their anxiety by imposing some form of understanding on the natural forces that make life tenuous. Religions abound for this purpose, and it is obvious that the sun and moon must have been our primary natural gods. I can believe that it was very comforting to these people to predict the risings and settings of the gods and even their eclipses. Fear would be allayed by confident predictability. There were no doubt great ceremonies celebrating the promised midsummer rising of the sun above the Heelstone, and those wizards who understood it all would, if given a bit of pomp and showmanship, be very powerful men indeed. There is also evidence of ceremonial cremations and burials on the site and around it. The entrance to the horseshoes opened to the northeast midsummer sunrise, so the natural direction for a temple, away from its entrance, was toward the southwest midwinter sunset. The deathlike ceremony of this cold, stark setting of the god would have been tempered by the promise of another midsummer rising. The religious images survive. Jesus's promise, "I am the resurrection and the life. He who believes in me will not perish," was doubtlessly proclaimed in another form by these prehistoric priests. And if the sun and moon were two visible gods, the unseen line of lunar nodes that allowed one to eclipse the other must have been a third more abstract and invisible god. Evidently the unseen god was even more powerful than the visible ones. Such a Trinity sounds familiar to us today.

After an hour of mindlessly staring at the stones, stepping over the Aubrey Holes and walking around the ditch, we sat down near the center. It was near midsummer, but I wasn't ambitious enough to await the next day's sunrise. Presently I noted that the early afternoon sun cast near my feet a shadow of a giant stone—Sarsen Stone number fifty-six. It was one of the two uprights of the central "Sunset Trilithon." It was the largest of the five trilithons, but one of the uprights and the stone lintel that capped the two uprights has fallen. The remaining stone fifty-six is the largest at Stonehenge, weighing about 50 tons. It is almost 30 feet long, and the 21 feet standing above ground was hand-worked by countless man-hours of pounding and scraping into a rectangular slab. My eyes shot up to catch the gleaming sun just aside the top of this memorable stone, and I suddenly felt like one of those dumb apemen who awake beside such a slab in the film *2001: A Space Odyssey*. I have often wondered if Clarke and Kubrick did not aptly use this stone as an inspiration for their model of a cosmic intelligence.

Through many years of hard work by literally thousands of scientists I believe that I have learned much more physically about the sun and moon than the Stonehenge people could have imagined. I have even had the good luck to talk to men who have walked the surface of the moon. Yet with each new fact, several new questions always occur, and the net effect is that the overall mystery continually deepens. Cosmology is the study of the universe. In a philosophical sense the studies reveal people's view of the universe rather than the universe itself. Their view of structure reflects them, the viewers; therefore, the study of cosmological thought is also the study of humanity. What could be more satisfying? If, as the poet has

The moon moves in front of the rising sun in this multiple-exposure sequence taken from Costa Rica on Christmas Eve, 1973. (Dennis DiCicco)

said, "The proper study of mankind is man," there is ample reason for a thorough scientific study of Stonehenge. Those stones stand as silent witnesses to an unwritten chapter in our intellectual development; and the sky was its source.

CHAPTER III
REVOLUTIONS

On my desk before me lies the preliminary program for the meetings of the International Astronomical Union in Poland in September, 1973. The IAU, as we call it, meets every three years to foster international cooperation in pursuing the goals of astronomy. The site is chosen, by an international council, to be near some past or present developments in astronomy. The only two I have previously attended, in Prague in 1967 and in Brighton in 1970, both provided lasting memories for me, but neither had the symbolic significance of the upcoming meeting. This one marks the 500th year of Nicholas Copernicus, who was born on February 19, 1473, in the Polish town of Torun. The earth has since then completed 500 revolutions of its 590-million-mile annual orbit around the sun at a speed of 67,000 miles per hour. After this immense journey of 295 billion miles, the town of Torun has returned to its same location relative to the sun. In midsummer of 1973 at Stonehenge the sun rose, exactly on schedule, for the 500th time from the tip of the heelstone since the summer following the birth of Copernicus. It is the thoughts of mankind that will not return to where they were, for Copernicus dealt in revolutions. The revolutions of heavenly bodies were his primary interest, but they led us to a revolution in human thought.

Of all the subjects under deliberation in Poland in September, 1973 the most interesting, to me, was the Symposium on Copernicus, planned for September 7 and 8 in Torun. The town is reputed to be rich in architectural relics. The old Town Hall, with the monument of Copernicus in front, and the church of St. John are considered among the finest examples of gothic style. The scientific program for the meeting will be concentrated in the new campus of the Nicolaus Copernicus University. From September 10 to 12, 1973 the scene shifts to Kracow, the historical capital of Poland and, in the past, the principal center of the nation's cultural life. It is also the richest of all Polish towns in historical and architectural monuments and relics over the past 1,000 years. Kracow is also the seat of the oldest Polish university, the Jagiellonian University; in the years 1491–1495 one of its students was Copernicus. Nearby lie the modern dread remains of Poland's darkest hour, Auschwitz and Birkenau, the largest of the Nazi concentration camps. The IAU Symposium in Kracow, "The Confrontation of Cosmological Theories with Observational Data," is chosen to reflect the revolution in

Statue of Copernicus outside the town hall in Torun, Poland
(*Sky and Telescope*)

thought that the cosmology of Copernicus thrust upon the world. I cannot repress an inner excitement at the symbolic meaning of this tribute.

In Houston Mayor Louie Welch proclaimed February 19 as "Copernicus Day," and various toasts were made by members of the Copernicus Society. It's unusual, and I have had to pause and reflect why this particular man was so important for us. One reason is that he is a symbol of the renaissance—not only of the historical sixteenth century of which he was so grand a part, but also of the renaissance in the minds of each of us when we overthrow an old and faulty explanation in favor of one based on fact and reason. The ideas of Ptolemy, the greatest ancient authority on geography and astronomy, had held sway for thirteen centuries. They were themselves great ideas because they explained the apparent motions of the heavenly bodies with admirable accuracy. The fundamental question on which Copernicus departed from Ptolemy was the cosmological status of the earth. Instead of regarding the earth to be the fixed center of the universe with all heavenly bodies moving around it, as it undeniably appeared, Copernicus imagined that the earth was just another planet, like Mercury, Venus, Mars, Jupiter, and Saturn, spinning daily about its own axis and orbiting the sun in one revolution per year. The radical thing was the contention that things aren't what they seem; that one cannot trust one's naive judgment. The sun does not set at night by sinking in the west and passing underneath the earth to rise again in the east—it only looks that way when we look at a fixed sun from a spinning earth. The seasons occur not because the sun migrates annually from north to south in its westerly passage, but because the tilted axis of the earth's daily rotation does not change direction as the earth moves in an annual orbit about the sun.

These ideas are hard enough for us today as we, in our childhood, retrace with parental guidance the basic fact of the heavens. Imagine the intellectual panic and the blind resistance of a stubborn mankind that had contented itself for thirteen centuries with Ptolemy's Earth-centered view. Why don't objects spin off the earth if it is indeed a spinning ball? Every farmer who ever handled pails of water knew about centrifugal force. The earth is obviously not whizzing along at high speed, or all loose objects would sail swiftly off in the opposite direction, and the wind would be fierce—or so it seems. The earth cannot be moving because a stone dropped from a tower falls straight down. (This objection was not actually based on a true fact, but towers were not high enough and measuring instruments fine enough to detect the apparent curvature caused by the rotation of the earth.) No wonder that in a time when funeral pyres occasionally burned witches and dissidents, Copernicus delayed publishing his book *De Revolutionibus Orbium Caelestium* until he was almost on his death bed. Copernicus died the same year, 1543, that his book was first published in Nuremberg. At that time it might have mattered little to inquisitors that Copernicus had diligently established arguments for his propositions.

When I was a boy I enjoyed the autumn celebration of the State Fair of Texas in Dallas. The midway of the fair boomed with traditional carnival activity ("Step right in, folks, and see the fat lady with two heads") and pleasure rides of great variety. It was a different world from parks like Disneyland and its many offspring, including Texas's own Six Flags Over Texas and Astroworld. For $1.00 I could expect to have quite an afternoon, especially with some free repeat rides I often got when traffic was low. There were two rides I especially enjoyed. I am reminded of one of them now; later I will return to the other. The double Ferris Wheel had a large outer wheel that turned relatively slowly. It provided a panoramic and slowly changing view of the midway. Within it was a smaller concentric inner wheel that completed its rotation in shorter time. The rides lasted a long time, and one of the fun things from the inner wheel was to fix one's sight on someone on the outer wheel. There was some fascination in how the inner wheel would, due to its faster rotational velocity, repeatedly pass the persons on the outer wheel. In so doing I always marveled at the very phenomenon that unlocked the solar system to Copernicus. Both wheels revolved in the same direction, and the person on the outer wheel usually seemed to be moving in that same direction as seen against the background of distant objects. The exception occurred when I overtook the person on the outer wheel. While I was passing him, he seemed for a short while to move in the opposite direction against the distant stationary background. I never really tired of that amusing reversal.

The planets of the solar system revolve in an easterly direction in their orbit about the sun. Therefore the planet Mars, whose orbit lies outside that of the earth, usually appears to move easterly through the distant stars. Copernicus concentrated on the fact that when Mars is at its brightest and is overhead near midnight, it always seems to reverse its direction and move westerly for a couple of months, whereafter it turns and moves east again. He realized that if the earth were a planet interior to Mars, then Mars would appear to move backwards for a couple of months while the earth was passing it. It is an explanation simple and profound, and it worked much more naturally than Ptolemy's epicycles. Copernicus made similar observations for all the planets. From them and from his hypothesis, he was able to place the planets in order of their increasing distance from the sun and was able to predict their apparent motions through the fixed stars with admirable accuracy. He was also able to obtain accurate values for the radius of the orbit of each planet and the period of its revolution (its "year"). With penetrating insight Copernicus had reduced a large variety of complicated observed motions to being only effects caused by the motion of the earth and the other planets. The actual system was beautifully simple, it only looked complicated when viewed from the earth, which is, after all, only one moving cog in the machinery of the solar system. This was a profound revolution in the study of astronomy, which would have otherwise remained an enigma. He gave the earth its true place in the cosmos.

It was hard for stubborn individuals to accept the principle that things may be much different than they appear to be. For many, it was also spiritually difficult to accept the fact that the earth is not the center of the universe, and is in fact only one planet among many orbiting one of the millions of stars observable with a small telescope. Religious issues were affected, and another great spirit of the renaissance, Martin Luther, exclaimed upon reading Copernicus: "Does this madman not know that it was the sun and not the earth that Joshua ordered to stand still?" The old system continued to be taught in the universities for another century. The tide changed very slowly in favor of Copernicus's ideas. A tribunal of the inquisi-

Constellation Orion is toward lower left as it might appear with good binoculars or a small telescope. The two brightest objects above it are the planets Saturn trailing Mars on their seasonal wanderings across the sky. The Pleiades star cluster is at the right above center. (Harvard College Observatory)

tion had declared in 1616 and 1633 that Copernicus's theory was heretical and condemned all books that maintained the motion of the earth. Yet the first page of an edition of Copernicus's book published in Amsterdam in 1617 reflects the true situation with a curious illustration. The heavens are being weighed against the earth on a simple balance, and they are winning; the earth is being deprived of its overestimated importance to the cosmos. Humanity's view of itself had to become more realistic as well.

No wonder that tributes continue. For its 1973 annual meeting the United States National Academy of Sciences specially commissioned a musical cantata, "Copernicus: Narrative and Credo," which had its first performance in the Academy auditorium on April 22. The composer, Leo Smit, is a native of Philadelphia, who then held the appointment of Composer-in-Residence (1972–1973) at the American Academy in Rome. The text was written by British cosmologist Sir Fred Hoyle, President of the Royal Astronomical Society and previous collaborator with Leo Smit. Their work is scored for narrator (a musically bold but appropriate step), mixed chorus, and a chamber ensemble of nine instruments. Musical sources for the work were drawn from the little known but highly developed music of medieval and renaissance Poland. The twelve sections of the work are: *Introduction* ("From the Dark Ages onward . . ."); *Birth; The Uncle; Cracow Student Song; Columbus West* ("Round . . . turn . . . spin . . . whirl . . ."); *Italian Madrigal* ("Non mi piango, non lamento"); *Sorrow and Solitude; The Teutonic War; Papal Disputation; The Book* ("de revolutionibus orbium caelestium"); *Death; Laudemus* ("So let us praise a man . . ."); and the concluding *Credo* ("I believe in one World . . .").

I cannot contemplate the celebration of Copernicus without thinking of the many times I walked through the great gate of Trinity College, Cambridge, and across its majestic old courtyard, because to me this story ends there. On my own pleasant way between Trinity Street and Queens Road, I sometimes stopped off in the Trinity Chapel to pay respects, so to speak, to Isaac Newton. He was a student in this fabled place, and his statue greets all who enter the chapel doors. Newton was born 99 years after Copernicus died, but his genius gave the Copernican theory a level of understanding and acceptance that it could not have had on the basis of observations alone. Indeed many universities taught the Copernican system as a convenient device for getting the right answer without fully acknowledging its physical reality. Newton explained planetary orbits with an argument that I always find appreciated in conversations. He imagined a cannon firing projectiles horizontally from the top of a single high mountain on an otherwise smooth earth. The projectile falls toward the surface of the earth (the "apple") while its horizontal muzzle velocity carries it away from the mountain. It therefore follows a curved path which hits the ground some distance from the base of the mountain. If the horizontal velocity from the cannon were increased, the projectile hits further away from the base of the mountain. As it is increased even more, the curved falling path follows the curvature of the surface of the earth, and *the object falls all the way around the earth!* Since the initial conditions are repeated when the projectile comes back to its initial firing point, it will remain in orbit indefinitely. This was actually accomplished on October 4, 1957, when the Sputnik 1 went into orbit around the

Statue of Newton in Trinity College Chapel, Cambridge, England

earth. Newton extended this concept to the falling of the planets around the sun. By expressing the argument in mathematical form, he was able to calculate the elliptic orbits determined by Kepler in an important refinement of Copernicus's circles.

Something very subtle and important had happened between the times of Copernicus and Newton. It was the development in human thought of the mathematical concepts of *functional analysis* and *space*. Copernicus used the same primitive mathematical tools of Ptolemy to describe the observed motions—he only changed the point of reference. Kepler envisioned a mathematical *background space* in which these things happen. He thought of the *path* of the planets rather than of some artificial orbits, and he stated clearly that *the path lies in a plane*. He successfully envisioned this path as being described by the manner in which the radius from the sun to the planet varied with angular position as the planet traces its path, whereas Copernicus had been content to describe the apparent position in time. This allowed Kepler to find a simple functional form for the path. He was even able to state clearly that this path has the properties that the radius vector sweeps out equal areas of its plane in equal times—a powerful mathematical statement—and, in a flash of synthesizing genius, that the squares of the periods (the planetary years) were proportional to the cubes of an important parameter of the functional analysis, the semi-major axis of the ellipse. The sky was contributing to the emergence of functional analysis! Newton carried this to its conclusion by developing the infinitesimal calculus to analyze his laws of motion and the paths they produced. One of the beautiful intellectual structures of the human mind was born, and its parents were the same as those of Stonehenge.

Newton showed that the motions of the Keplerian system were exactly those one would expect with the law of universal gravitation. These brilliant arguments demolished the last waves of intelligent resistance to the Copernican revolution, and they earned for Newton an almost unparalled homage in the history of intellectual thought. I too gaze at his statue in the chapel as if I am establishing a magical rapport with the old rascal—an arrogant, irascible, lonely, and even misanthropic man. I never succeeded, but I have tried again and again just the same.

CHAPTER IV
THE BRIGHT NIGHT SKY

There is a small train station near Place Chauderon in Lausanne where one can take the Lausanne–Echallens–Bercher line. This is a one-car electric train, making half a dozen stops at various small places in its 25-kilometer route to the end of the line in Bercher after it pulls out amusingly into the face of oncoming traffic on Avenue d'Echallens. Just 7.5 kilometers north and slightly east of Lausanne it passes through Cheseaux, a small Swiss village of roughly 400 inhabitants. The village has an elevated position relative to Lausanne, lying on a plateau tailing off from the west side of the Jura mountains on the water routes down to Lac Léman (Lake Geneva). The little river Mebre flows unimportantly by.

Within fifteen minutes I walked every street through the village. Despite having a bank, a school, and a police station and despite lying on (and being virtually destroyed by)the Lausanne—Yverdon highway crossed by a major road to Geneva, Cheseaux doesn't seem to me to be part of the twentieth century. Maybe that's because I'm aware of its past. I saw only two buildings of outstanding architectural interest—a beautiful eighteenth-century château with many gardens and adjoining buildings and a more modest multipurpose farmhouse with typical Savoyard characteristics, probably dating from the sixteenth century. Both buildings belonged at one time to the distinguished family Loÿs, the head of which was, by tradition lasting over two centuries, considered the local nobleman. The older building is surrounded by high stone walls and it rises proudly but mysteriously above the other village houses to the left of the post office, about halfway from the Yverdon highway to the railroad tracks. It is a curious structure with elevated doors suggesting that associated structures no longer exist. In a tragic attempt to convert this sixteenth-century house to a modern multiunit dwelling, the front half of the building has been completely refinished. In the back half, with broken or bricked-in windows, ancient wooden beams, and cracking stone walls lives an elderly eccentric who was amazed that I could be interested in the place. Neighbors watched with condescending amusement as I took photographs.

The estate was purchased in 1557 by the family Loÿs from the government of Bern, and they owned it until it was sold in 1769, again to the Bern government. The grand château, which stands at the edge of the village on the Route de Genève, is an eighteenth-century country house that was sold along with the rest of the estate. In keeping with the tradition of the times, it was probably remodeled after the sale. A magnificent structure with pink walls, it can be found in pictorial collections of fine Swiss châteaux. It seemed a shame to me that the residents seemed unaware that this estate was the residence of an ill-fated genius—a prototype of the prophet without honor.

I had come to Lausanne on holiday, so it was curious that I now found myself in this little village in search of an ancient cosmologist. Shortly before the visit I had accidentally stumbled on an article by the famous modern astronomer, Struve, in which he claimed that the real discoverer of the mystery of the darkness of the night sky had been an astronomer of Lausanne. That caught my eye because my interest in the profundity of that simple observation had been growing. By rummaging about in the University of Lausanne Library and the Archives Cantonales I discovered that the astronomer in question had actually lived in Cheseaux rather than in Lausanne. Responses to my questions had led me to this small train line. During the ride there my imagination skipped back two and a half centuries, trying to imagine how some of humanity's greatest concepts could have arisen in that world of simple farming peasants and occasional noblemen. Now I searched the village in vain for any remembrance that anyone special ever lived there.

Jean-Philippe Loÿs de Cheseaux was born in the spring of 1718, almost certainly in the eighteenth-century château. He was a child prodigy despite a sickly constitution. He was taught solely by his grandfather, a distinguished scholar, with special emphasis on science, theology, history, and languages. He knew Latin, Greek, Hebrew, Arabic, Aramaic, and English. Despite being very ill through part of his teenage years, Loÿs was a precocious physicist who at the age of seventeen published a treatise on physics, and he became a famous astronomer. In 1736, at the age of eighteen, he constructed the first astronomical observatory in eastern Switzerland. The observatory is reputed to have been in a tower about 60 feet in height, which was called "Dixmes" by the inhabitants of Cheseaux. It no longer exists and, incredibly, its exact location is apparently unknown, although it was definitely near or in the village. The only reference I could find to it reported that in about 1852 it was changed into a habitable residence that was destroyed early in this century.

In 1743 and 1744 Loÿs de Cheseaux made from his observatory outstanding observations of two comets, one of which had six spectacular tails. His published book on these findings, *Traité de la comète qui a paru en décembre 1743*, won him considerable fame and he was invited to the French Academy of Sciences in Paris. It is an appendix to that book that interests me because it ensures his prominent place in the history of man's thought about the universe. Entitled "Sur la force de la Lumière et sa propagation dans l'Ether, et sur la distance des Etoiles fixes," that appendix addressed itself to the question of the darkness of the night sky and the distances of the fixed stars.

Why *is* the night sky dark? Knowing from our earliest years that the night sky is dark, it seems a strange question. We accept the fact so completely, so instinctively, so ingrained in our thoughts that it almost seems to require no explanation. It had long been recognized that the darkness is due to the absence of the sun, and even before the Copernican revolution it was known that this daily absence is due to the daily rotation of the earth, so that night occurs when the earth's rotation is such that the sun is on the opposite side of the earth from us. Loÿs de Cheseaux accepted that correct explanation, but he went on to inquire what it means that the night sky is not made bright by the other stars. That was his big step, and if it doesn't seem to us to have been a big step in 1743, perhaps we should ask ourselves a question.

Why don't the other stars make the night sky bright? About the best response we can immediately make is to call up our vague notions that there are too few of them, and they're too far away. That answer, as far as it goes, is undoubtedly correct, but Cheseaux realized that it had some important implication for the structure of the universe. At that time it was already generally recognized that the sun was just another star, although one of special importance to us because the earth is in orbit around it. The star-filled heavens seemed to suggest that the universe goes on without end. Indeed religion still played a big role in the lives of educated men, Cheseaux included, and it would have seemed unduly doubting to place limits on God by suggesting that the universe of His creation exists only near to us. Thus it was commonly assumed that the universe is infinite—that its star-filled space goes on forever. This assumption followed the ideas of Giordano Bruno, the revolutionary sixteenth-century philosopher. In a letter dated December 10, 1692, Sir Isaac Newton showed that he too believed the assumption to be correct. It was Cheseaux's brilliant accomplishment to realize that if that were indeed so, as men said, the night sky should be bright unless something else causes it to be dark.

It is of considerable interest to reconstruct what Loÿs de Cheseaux did. This first step was in itself a great accomplishment—the measurement of the distances to the fixed stars, in particular to the so-called first-magnitude stars which are quite bright even to the unaided eye. At this early time no telescope had sufficient pointing accuracy to determine the distances of nearby stars by the modern parallax technique, which works as follows. At times six months apart the earth lies on opposite sides of the sun in its annual orbit of the sun. Between these two times, therefore, the position of the earth changes by twice the distance from the earth to the sun (or two astronomical units, as the earth-sun distance is called)—about 3×10^8 kilometers, or 300 million kilometers. When the fixed stars are viewed from the two points separated by this amount, which though large to us is nonetheless very small compared to the distance to other stars, a funny thing happens; namely, the positions of the nearest visible stars seem to have shifted with respect to the positions of the more distant stars. (To visualize this effect one needs only to recall how from a moving automobile the nearby scenery appears to move past the scenery on the distant horizon.) The difficulty is that the apparent angular displacement of the nearest stars is only about one second of arc (1/60th of a minute of arc, which is itself the apparent angular size of a grape from the opposite end of a football field). Cheseaux's instrument could certainly not point that accurately, and in fact this method was not technologically possible until 1838. Therefore Loÿs de Cheseaux had to measure the distances to stars by some other technique.

It seemed very natural to Cheseaux to assume that the other stars have the same luminosity as the sun, but that they appear less bright because they are so far away. Today we know that because this assumption is not exactly correct it is only a rough way to estimate the stellar distances. Every physicist of that time knew that the light intensity received from a source of fixed luminosity varies inversely with the square of the distance to the object; that is, stars twice as far away should be ¼ as bright, those ten times farther away 1/100 as bright, etc. This assumption allowed Cheseaux to estimate the relative distances to different stars, but it did not give the absolute distance to any. To obtain the absolute distance to any star by this technique, he had to compare the brightness of the star to the brightness of the only star whose distance he could determine—the sun! A little thought will show that this must be a difficult task. Not only is the star visible only at night and the sun visible only by day, but they differ so enormously in brightness that there is no elementary and natural way to compare them. The ingenious solution used by Loÿs de Cheseaux was to compare the apparent brightness of the first-magnitude stars with that of the planet Mars. They were nearly enough equal that the comparison could be easily achieved. The distance to Mars was reasonably well known, so the only remaining link needed was the comparison of the brightness of Mars to the brightness of the sun. This comparison also could not be done directly, but it could be calculated in the following way. Because Mars is a planet it has no luminous power, but instead only reflects the light from the sun. Thus the luminosity of Mars is the same small fraction of the luminosity of the sun as the ratio of the area of the Martian disk is to the total area of the surface of the sphere containing the orbit of Mars. Carrying out this entire procedure, Loÿs de Cheseaux found that the distance to the first-magnitude stars was 240,000 Astronomical Units (240,000 times the earth-sun distance). Because the distance to any star is so much greater than the distance to the sun, it is more convenient to use the light year, which is the distance traveled by a light ray in one year. Cheseaux's distance for the first-magnitude stars was 3.7 light years, which is in itself an amazing accomplishment considering that the correct value using his argument is now known to be 5.6 light years. In actual fact, the distance to the nearest first-magnitude star, Altair, is known to be 15 light years by direct measurements using the parallax method. The discrepancy occurs because the first-magnitude stars like Altair are actually more luminous than the sun. Nonetheless, Cheseaux's was the best measurement of stellar distance that had been made at that time.

Men often rest after such an accomplishment; or sometimes they are content to only polish their result. Fortunately it is occasionally true that new results will stimulate the minds of extraordinary men to make an intellectual leap into a world of totally new thought. In this sense Cheseaux stands in line with Bruno, Galileo, and Newton because he next made a calculation that is in many respects the prototype of all cosmological-astronomical calculations. Although the world was not aware of it, its conception of our universe was to be forever changed by it.

Jean-Philippe Loÿs de Cheseaux had a forceful mastery of geometry. His fascination with it pervades his studies, large and small, and apparently it caused him to consider the apparent angular diameter the sun would have were it at the distance of the first-magnitude stars. He knew that the diameter of the sun was almost 31 minutes of arc as viewed from the earth one Astronomical Unit away. He calculated that at a distance of 240,000 Astronomical Units the sun's diameter would be a small (but nonvanishing) $31' \div 240{,}000$ minutes of arc, or 0.008 seconds of arc. He estimated that there exist twelve stars of first-magnitude, and from that he concluded that the first-magnitude stars have a total disk area equal to $\frac{1}{4 \times 10^9}$ times the area of the solar disk. Not stopping there, he reasoned that the same small fraction of the solar disk would be occupied by stars four times fainter than the first-magnitude stars because, from the inverse-square law, they will lie on a sphere having twice the radius of that to the first-magnitude stars, and such a sphere has an area four times greater and hence should contain four times as many stars; that is, the surface of a sphere twice as far away contains four times as many stars, each of which has an apparent angular area ¼th as great. The next sphere three times farther away has nine times as many stars of one-ninth the area so that they too occupy the same fraction, $\frac{1}{4\times10^9}$ of the solar disk. Obviously when 4×10^9 such spheres are considered, the sky would be covered by a star area equal to the area of the solar disk. That's as bright as day! Cheseaux went on to estimate that the sun's area goes 92,000 times into one hemisphere of the heavens, so that when $(4\times10^9)(9.2\times10^4) = 3.7\times10^{14}$ such spheres are considered, the entire sky would be as bright as the surface of a star! Using his distance of 3.7 light years as the radius of the first sphere, Cheseaux argued that bright stars must not be visible to distances as great as 10^{15} (one million billion) light years. Either there *are* no such stars or something prevents their light from ever reaching the earth. That conclusion, simple but powerful, stands today.

Three solutions may have occurred to Loÿs de Cheseaux, although his published appendix addressed only the third (incorrect) alternative. The first solution would be that there are no far away stars because the universe ends nearby. The rejection of this possibility because of leading scientific and religious opinion was the basic cause of the paradox. To question the extent of the universe seemed tantamount to questioning God's omnipotence. Modern astronomy continues to reject this solution to the dilemma—but on scientific grounds. As each new instrument has looked deeper and deeper into the vastness above, the ever fainter galaxies of stars continue to appear. And their distant density seems to be the same in all directions. It violates our Copernican sophistication to assume that we are near the center of a large but limited universe. Nonetheless, I must admit that although I also do not believe this solution is correct, I know of no unequivocal proof that it is wrong. But a confined universe is out of touch with all modern thought and observations, and, instead of being simpler, a universe with boundaries raises more philosophical and physical problems than it solves. I can recall baffling myself as a boy over "what lies beyond the edge." Like Loÿs de Cheseaux, we reject the idea, for different reasons perhaps, and with some mental insecurity perhaps, but we do reject it.

I am surprised that Loÿs de Cheseaux was not attracted to a second solution to the problem—finiteness in time instead of space.

Ever since the Danish astronomer Ole Roemer discovered in 1676 that the apparent time of Jupiter's eclipse of its own satellites seemed to depend on the distance from earth to Jupiter, it had been generally believed that light travel is not instantaneous. To account for the shifts it was required that the time required for light to travel one Astronomical Unit be 8 minutes and 19 seconds. Today we know that speed for light to be correct by actual measurement, and even without proof Cheseaux almost surely accepted the explanation. Yet if he accepted it, it would follow that the light we are receiving from stars on the sphere 4×10^9 light years away, for example, would necessarily have been emitted 4 billion years ago! This long time was in the eighteenth century much greater than the commonly accepted age of the earth. The biblical account seems to place the creation near 6,000 years ago, and Cheseaux accepted a literal and strict interpretation of the Bible. Two of his more interesting religious efforts were an attempt to obtain the exact date of Christ's crucifixion by assuming that the darkness on Golgotha was due to an eclipse of the sun, and a prediction from his study of Daniel and the other prophets that the second coming of Christ would occur in 1749. When that year passed uneventfully, Cheseaux published an apology with the observation that matters of moral significance were not calculable by scientific techniques. But his entire life showed a devotion to traditional theology, and it seems likely that he accepted the notion of the creation. If that be the case, the universe is young, even though infinite, and no extremely distant stars could be seen because their light would have to have been emitted before God's command, "Let there be light," or it could not yet have reached us. Curiously enough this explanation, which Cheseaux either did not think of or rejected, plays an important part of most modern explanations of the darkness. It is commonly believed that stars did not exist before about 10 billion years ago, although many scientists are still skeptical of the correctness of this majority opinion. But if it is correct, then galaxies more distant than 10 billion light years in our infinite universe cannot be seen because their starlight has not yet arrived. Ordinarily one might expect that no cause outside the Deity Himself could contrive to have all galaxies of stars form simultaneously, especially considering that they have no way in principle (except at speeds greater than that of light) of communicating their births. But the modern models of evolving universes, or "Big Bang" cosmologies, do allow this synchronization in a fairly natural way —so perhaps it is correct.

Loÿs de Cheseaux actually chose a third solution. In "Sur la force de la Lumière et sa propagation dans l'Ether" he proposed that the ether, or space through which light travels, absorbs light slightly. To account for the observed illumination of 2×10^{-4} footcandles, he argued that the interstellar space needed to be not totally transparent but 3×10^{17} times more transparent than water. By quantitative comparison the light of the first-magnitude star Aldeberan would be degraded by the equivalent of only 4 inches of water. In this way the effect of the most distant stars could be eliminated. Interestingly enough, Loÿs de Cheseaux's arguments were repeated eighty years later by the German astronomer Wilhelm Olbers, whose only purpose was to show that the required absorption could be provided by the interstellar gas and dust, and yet the argument is now often referred to as "Olbers's Paradox." This is a historical injustice that should be put right. Nonetheless, both men overlooked a simple fact: Any medium irradiated by a sky as bright as the surface of a star must soon heat up to that same temperature, whereupon it only reradiates all it absorbs. Thus this explanation fails.

Nonetheless the world will remember Jean-Philippe Loÿs de Cheseaux, an ailing genius who saw the way to a profound question. We are right to marvel that the sky is dark, no matter how accustomed we become to the commonplace. At the height of his fame as a young man, Loÿs de Cheseaux refused the prestigious position of director of the St. Petersburg Observatory, preferring to remain in Cheseaux due to his weak health. In 1751, while delivering an address to the French Academy of Science in Paris, he became again seriously ill. He died there shortly later, at the age of thirty-three, ending a brief lifetime of mental strength and physical weakness. It was five years before the birth of Mozart. Cheseaux's scientific instruments have never been discovered, despite numerous searches in Paris and at the homes of his descendants. Perhaps they were transported to Bern when his brother, Charles-Louis Loÿs de Cheseaux, sold the family estate in 1769, just eighteen years after Jean-Philippe's death. At this juncture, it appears unlikely that they will ever be found. He was like the comets he observed, a bright flash speeding by unnoticed by humanity. Nothing much remains but his papers, his books, and his portrait. Painted in 1746, when

Loÿs de Cheseaux was twenty-eight years old, by the artist J. P. Henchoz, it hangs in the Salle du Sénat of the Palais de Ruminé in Lausanne. His globe is placed beside him, and Newton's *Principia* stands on the bookshelf behind him.

A second portrait, painted by Annette from my photo of the original, hangs in our home. His gaze reminds me that brilliant and gentle spirits wrestled with cosmology long before the turmoil of the modern world. I have come to know his eyes as if they are still real. They almost haunt me, and staring into them I shudder as I realize that a single life is but a fleeting second. I breathe a quiet thanks for another day and resolve to use it well.

Jean-Philippe Loÿs de Cheseaux, "Sur la Force de la Lumière et sa propogation dans l'Ether, et sur la distance des Etoiles fixes" (1744)*

* The excerpt of this essay was translated from the French by Annette Clayton.

J. P. Loÿs de Cheseaux, painted in 1746 by J. P. Henchoz (Bibliotèque Cantonale et Universitaire, Lausanne)

It is a demonstrated proposition of optics that if all the fixed stars were equal and similar to the sun, so that placed at the same distance they would have the same apparent angular diameter and equal brightness of light, or that they would send to us the same quanity of light, it is, I say, demonstrated that the quantity of light that each of them placed at any distance from the earth would send to us would be to that of the sun in direct ratio to the square of its apparent diameter to the square of that of the sun, or, equivalently, to the inverse ratio of the square of its distance to the square of that of the sun. Imagining now the whole star-filled universe divided into spherical concentric shells, and with a density almost constant, and supposing the number of stars contained in each shell to be roughly proportional to the surface of this shell, that is to the square of the distance from the sun taken as the center of the entire starlit universe, and finally taking the true diameter of each star to be almost equal to that of the sun, as I have supposed from the very beginning, one finds the quantity of light sent to us by the stars of each shell to be proportional to the sum of the squares of their apparent diameters, which means proportional to the number of stars of each shell multiplied by the square of the apparent diameter of one of them, or, by what I have said, proportional to the square of the distance of each shell divided by this same square. Therefore the quantity of light is always the same for each of the shells, and each will add to the quantity of light we receive a fraction of the sun's light equal to the ratio of the square of the distance to the shell of first-magnitude stars to the square of the distance to the sun from the earth times the number of stars in the first shell, that means in the ratio 1 to 4,000,000,000. From that conclusion it follows that if the star-filled universe is infinite, or only greater than the distance to the first-magnitude shell by the factor 760,000,000,000,000, each point of the sky would appear as luminous to us as one point of the sun because the light we would receive from the two celestial hemispheres would be 91,850 times greater than that which we receive from the sun. The enormous difference between this conclusion and experience lets one see that the sphere of fixed stars not only is not infinite but also that it is uncomparably less than the vastness I have supposed for it, or that the strength of light decreases in greater proportion than the inverse ratio of the squares of the distances. This last supposition is probably enough; it demands only that the star-filled space is filled with any fluid capable of intercepting, be it as little as it wants to be, the light. If this fluid were 330,000,000,000,000,000 times more transparent than water, it would be enough to dim the intensity of light by a thirty-third part during its passage through each shell and to absorb by degrees the entire light from stars beyond our neighborhood to the point of reducing the light of a whole hemisphere to the 430,000,000th part of the amount of light we receive from the sun, or to an amount of light only thirty-three times greater than that which we receive from the dark globe of the new moon illuminated by the earth.

Without doubt one will judge that the numbers that I set up are to be taken as conjectures; this is true, but these conjectures are not arbitrary—least of all the distance to the stars of first-magnitude that I have placed about 240,000 times farther than the sun.

CHAPTER V

A DECADE TO REMEMBER: AFTERTHOUGHTS ON LOŸS DE CHESEAUX

The decade of the roaring twenties is a never to be forgotten one. Its astronomical exuberance was on my mind when, in October 1972, I took Annette for her first visit to Mt. Wilson. After we turned upward from La Canada onto the Angeles Crest Highway, the terrible pollution of the air of the Los Angeles basin became increasingly evident as we rose above it. We never became free of it because it extended to the top of Mt. Wilson, ending forever the clear view that had given mankind one of its most thrilling moments. While climbing the winding road in line with all the other traffic, my mind kept skipping back seventy years, trying to imagine mule teams on difficult two-day ascents taking astronomical equipment from Pasadena to the top of Mt. Wilson. It had been necessary to clear and grade paths for the special wagons that the mules pulled up and over the rough ground. To look at the majestic domes around the 100-inch and 60-inch telescopes on top, one marvels that it was done at all. It had been a labor of love supported by the Carnegie Institution of Washington.

In the 1910s and 1920s astronomers exclaimed lyrically of the clear, black, and noiseless nights far from the civilization below, while they saw surprise after surprise in their outward looks at the universe. George Ellery Hale and Edwin Powell Hubble will, in particular, never be forgotten for their key roles during those years. Hale was a great astronomer of the sun, and his organizational and fund-raising talents were in large part responsible for not only Mt. Wilson but also for the 200-inch Mt. Palomar telescope to be con-

Airplane view of Mt. Wilson Observatory (Hale Observatories)

structed decades later. The two observatories are now named The Hale Observatories, with offices in Pasadena near the foot of Mt. Wilson, where they are operated in close conjunction with the astronomy department of the California Institute of Technology. Edwin Hubble was, like Copernicus, Loÿs de Cheseaux, and Kepler, the astronomer of destiny, whose instincts would utilize the new instruments in their moments of great discovery. His demonstrable proof of the existence of a universe of galaxies outside our

The 100-inch telescope on Mt. Wilson (Hale Observatories)

own vastly changed man's philosophic outlook, and in so doing considerably altered the details of Cheseaux's argument.

Although Cheseaux's reasoning was correct in its principle, it is numerically inappropriate. Careful counting of stars shows that their density everywhere in space is not the same as it is observed to be near to the solar system. In looking over the whole sky one sees only twelve stars brighter than first-magnitude, whereas about 3,000 stars brighter than sixth-magnitude can be seen. Modern calibrations of this ancient magnitude scale of visual brightness shows that a difference of five magnitudes to the eye corresponds to a factor of 100 in actual brightness. Assuming stars to be equally bright except for their differing distances, a difference of five magnitudes also corresponds, therefore, to a factor of ten in distance. That is, the average sixth-magnitude star is ten times more distant than the average first-magnitude star. From this one can easily calculate that, if the density of stars were uniform, the number of stars brighter than sixth-magnitude should be 1,000 times the number of stars brighter than first-magnitude. The respective numbers should be proportional to the volume of a sphere whose radius is the distance to the stars in question. Since the radii stand in the ratio 10 to 1, the volumes are in the ratio 1,000 to 1. The assumption fails, however, because the respective numbers brighter than first- and sixth-magnitude are 12 and 3,000.

The unaided eye can see only about 6,000 stars, but modest telescopes and photographs have allowed the star counts to be made to dimmer magnitudes. The comparison can therefore be continued. The number brighter than eleventh-magnitude should, by the same reasoning, be 1,000 times more numerous than the number brighter than sixth-magnitude. The photographed number is only about 200 times greater, however; the actual number brighter than eleventh-magnitude is only 700,000 rather than the 3,000,000 one might anticipate from the product of $3{,}000 \times 1{,}000$. By photographing to five magnitudes dimmer yet, one sees that the number of stars is growing even more slowly, as if we were running out of stars. It is also noticeable that the number of stars grows more nearly as expected in the direction of the Milky Way, whereas it hardly grows at all in directions away from the Milky Way.

In our time the explanation of this is known even to many high school students. The Milky Way is a relatively thin flat disk of stars, and we are a part of it. Although there are about 100 billion stars in this disk, we see only a fraction of them because interstellar pollution blocks the view of the most distant when looking through the flat part of the disk. When looking perpendicularly out of the disk, however, telescopes see all the stars—and their numbers end. Our galaxy, the Milky Way, has an edge! It is as if one were to place light bulbs on poles of differing height spaced a few meters apart on a small Pacific island. The upward view would reveal no light bulb beyond the highest poles. The view with binoculars along the island would show increasing numbers of dimmer light bulbs until our sight reached the shore of the island. An "island universe" it was called, and scientists and philosophers alike found it an acceptable, if slightly strange, universe.

If that's all there were to the story, the "Cheseaux–Olbers Paradox" would be immediately solved. The night sky would be dark simply because the stars end. Personally, I am glad the question did not happen to have such a trivial disposition. When the 60-inch reflector was completed on Mt. Wilson about 1913, the early observations of spiral patches like M31 in Andromeda immediately suggested that they too were composed of stars—very distant stars. A photograph in 1917 showed a new star in an external galaxy where earlier photographs showed none, and it was immediately suspected of being a bright nova explosion. Subsequent photographs with the 100-inch reflector revealed many individual stars. The fuzzy white nebulae could be clearly seen to be giant galaxies of stars like our own Milky Way. Instead of a universe full of stars, we appear to live in a universe full of galaxies. Think, if you will, of the Pacific islander alone for all time with his tribe on their island, carefully turning his newly constructed telescope to the faint patch of light he had seen on the horizon beyond his shores with his binoculars, and seeing millions of very dim light bulbs clustered together on millions of closely spaced poles. That would have been a night to remember!

Placing ourselves in the situation of the hypothetical Pacific islander we realize that one of the most interesting questions would be the distance to the newly discovered island. There are at least four ways the islander could estimate its distance. One way would be to carefully measure the direction to the new island from opposite sides of his own island. The small angular difference of its apparent position would tell how far away it is compared to the size of his own island. Such astronomical surveying can measure the distances to nearby stars in our galaxy, but it proves too difficult for other galaxies because they are so far away that the angular change of direction is too small to measure. A second technique would be to assume that the distance between light bulbs there is the same as on his own island, in which case the apparent angular separation between light bulbs would show how far away the island is. This again proves too difficult for galaxies because their stars are so irregularly spaced and hard to resolve. In desperation he might observe the angular size of the entire island as shown by the light bulbs, in which case he could estimate its distance if he knew that the absolute size of the island was the same as his own. Unfortunately, galaxies come in widely different shapes and sizes, just as islands do, so this technique is unreliable.

The fourth and best technique is to estimate the island's distance from how bright its light bulbs appear to be. This was Cheseaux's method for estimating the distance to the first-magnitude stars in our galaxy. This method works for other galaxies, but its history has proved to be treacherous. The problem is that all stars are not equally bright. To understand the problem one must imagine that some of the light bulbs are like those on a small pocket flashlight whereas others are like those used to illuminate runways at major airports. Weak stars like the sun are not visible in other galaxies because they are too distant. Some stars are blue, some are yellow like the sun, and some are red. Thus one needs some way of knowing the absolute brightness of the stars. Nature has been kind to us here because one class of very bright stars, the Cepheids, are not constant but brighten regularly with periods of several days. Conveniently for us, they declare their own brightness because the brighter Cepheids have longer periods of pulsation. One with a 30-day period is known to be about three times brighter than one with a 3-day period. Their slow and regular pulsation can be seen at large distances because they are typically about thousands of times brighter than the sun. During 1923–1924 Edwin Hubble analyzed with the new 100-inch telescope on Mt. Wilson the light curves of the Cepheids he had been able to resolve in the nearby Andromeda galaxy. At a Washington, D.C. meeting of the American Association for the Advancement of Science, on January 1, 1925, he announced to the astonished world that the Andromeda nebula is 2 million light years away! That is, even the nearest light galaxy is so distant that light must travel for 2 million years in its journey from M31 to us.

Spiral galaxy, Messier 51, in the direction of the constellation Canes Venatici. From a great distance our own Milky Way Galaxy probably looks much like this one. (Hale Observatories)

However, the entire galaxy is quite bright, visible with binoculars, being about ten billion times more luminous than the sun. In terms of the island analogy, the galaxies would be about one mile in diameter and situated 30 miles apart. On one of them the sun is an invisible dot, about 1/10,000 of a centimeter from an even smaller planet Jupiter. The entire solar system is smaller than a grain of sand on its beach. As fantastic as these numbers seem, it is even more fantastic to realize that the big telescopes show the universe to have literally countless numbers of such galaxies.

Had Cheseaux known all this he could have described his argument in modern form: If bright galaxies fill unending space with uniform density, then either the night sky must be blazingly bright or some effect prevents the light of distant galaxies from reaching us. Take your choice, but take it carefully and with due regard to the other facts known about the universe.

Time has dimmed the usefulness of these telescopes somewhat. The air pollution of the basin now surrounds their white domes, and the reflected city lights brighten the night sky. On Sundays the line of motorists driving up the mountain resembles a line of foraging ants. They come not only seeking relief from the city below, but also to share in some small way the historic discoveries that were made here. Like pilgrims they climb the stairs to view the 100-inch telescope mount, a modern sculpture of huge old gears behind a viewing glass. Each person thinks his silent thought about the universe and departs. On good nights these telescopes still seek new secrets as best they can in the now inferior environment, while the huge old pine trees stand as they did half a century ago in the pristine stillness. It was a decade to remember.

PUNTING ON THE CAM

Stuck pole! Shall I let go
Or can I risk just one more tug?
Endless decisions confront me here,
Where new thought blooms.
Gliding punts, skirting sculls,
Yet my mind will not escape
Imagined ancient unseen eyes
Peering from those walls of stone.
Did Newton too struggle so
With such confounding pleasantry,
Or did that perspicacious vision
Spring only from the baser torments?
Etheldreda! and monks of Ely,
Did harsh fates and troubled times
Lead you and all we others after
To struggle here in our own ways—
You with life and death and trade
And I with simple navigation,
Perspiring as I thrust again
To reach my pint of ale?

Clayton

CHAPTER VI
WHY AREN'T THEY ALL GONE?

In Cambridge, on a pleasant afternoon after leaving the Cavendish Laboratory, it was a temptation to punt along the river, but when Annette would meet me it was irresistible. The punt is an integral part of Cambridge, itself one of the jewels in mankind's crown.

I went to the Cavendish each Wednesday afternoon to attend the research conference of Professor Ryle's radio astronomy group. Ryle and Hewish would be there, as would Shakeshaft, Longair, and other staff and research students. In these hallowed halls of physics, reeking with the memories of great innovators and Nobel Prize winners, a very modern subject is under study—radio astronomy. James Clerk Maxwell, one of the most eminent of the past Cavendish Professors, would, I presume, have loved seeing the electromagnetic waves predicted by his elegant theory being detected from extraterrestrial sources. Mostly the group discussed new research papers about radio galaxies or pulsars, which are under study there. I was not expert in these matters, but in 1971 while I was on sabbatical leave from Rice University in Houston I was trying to learn more about them.

Sir Martin Ryle and Anthony Hewish, the radio astronomy professors of the research group, have been awarded equally the 1974 Nobel Prize in Physics. It is a great event in the world of astronomy because this is the first Nobel Prize awarded specifically for astronomical observations, although both Bethe and Alfven before them were cited for contributions to astrophysics. In a broader sense, this 1974 award is a tribute to the whole world of radio astronomy, a research area that vastly enriches knowledge of the universe and revolutionizes the traditional boundaries of astronomy.

Martin Ryle had just joined the Cambridge research group when the Second World War erupted. These were the infancy days of radio astronomy, a field which was only discovered in 1932 by K. G. Jansky. Ryle, like Hoyle and other living great men of British astronomy, worked on aspects of radar development for the defense effort. Back in Cambridge in 1945 he saw correctly that the new radar techniques opened the gates to a new astronomy. In particular, Ryle developed ways of making an array of small antennas have the pointing accuracy of a huge finely finished radio disk. The latter approach was also developed beautifully in Britain, by Sir Bernard Lovell at Jodrell Bank near Manchester. Other post-Second World War radio astronomy efforts were begun in Australia, the United States, the Netherlands, and France. Ryle himself is quiet and gracious, though very determined and even a trifle stubborn. Though it has little to do with his Nobel Prize, Ryle gained international notoriety for his early attacks on Hoyle's steady-state universe, some of which were premature and some of which are standing the test of time. The Cambridge scene was always intensified by rumors of rivalry between these great astronomers—but more of that later.

Anthony Hewish was awarded his share of the Nobel Prize for his important role in the discovery of the pulsars. Hewish had joined Ryle's research group in Cambridge in 1946, and his interests in the transient effects of the earth's ionosphere on radio waves ultimately placed him with the capability to detect the rapid variations from the pulsars. With research student Jocelyn Bell, he identified those extraterrestrial signals in 1967, although they took time in secrecy before announcing their discovery that rocked the world. Subsequent investigations around the world have confirmed that the pulsars are rapidly spinning neutron stars, stars so dense that a mass equal to the earth's occupies a cube the size of a football field! The mass of the entire sun has been squeezed to a sphere a mere 6 miles in radius. The entire matter is similar to that at the nucleus of an atom. It was indeed an astonishing discovery—all the more so because it was unpremeditated. The conceptual problems associated with the pulsars have stretched the minds of men everywhere, and both astronomy and physics are much the richer for it.

Annette was waiting for me on Free School Lane just outside the Cavendish Gates. After a short walk along Botolph's Lane, where Willy Fowler told me the Burbidge's once lived, and down Silver Street, we reached the Scudamore boat docks by the mill. From there we could punt along the river past the backs of the old colleges that have in part made Cambridge what it is—Queens College, Kings College, Clare, and Trinity, where Newton studied, St. Johns, where Hoyle was a Fellow, and finally Magdalene College, by the old traffic bridge. There, at another Scudamore boat ramp we woud end that incomparable journey, only 50 yards from the house where we lived on Thompson's Lane. Often I lay back and acted the philosopher while Annette punted us along.

Cambridge is one of the places with a long, continuing history of great cosmology, dating back to the seventeenth century when Newton's personal vision changed mankind's. The science of cosmology often moves forward by drawing great conclusions from commonplace facts, as Loÿs de Cheseaux and Newton both demonstrated. One of the first great cosmologists of this century was Ernest Rutherford, Cavendish Professor from the year 1919 until his death in 1937, Nobel Laureate in 1908, and, along with Einstein, the intellectual leader of the world of physics for the first three decades of this century. Rutherford's birth 100 years earlier in 1871 was being commemorated during our year there. He had first come to Cambridge in 1895 as a research student in the Cavendish. Curiously enough, Rutherford's first work there was also on radio waves. One fascinating feature of his radio physics was the way he was able to increase the distance between sender and receiver. He had a receiver installed in the top floor of the Cavendish Laboratory and detected signals from Park Parade, more than half a mile away. Today the radio astronomers detect objects so distant that billions of years are required for the radio waves to reach the earth. Rutherford did not come to Cambridge destined to work on radio physics, however, for he was to be the discoverer of the structure of the atom and the subatomic laws of radioactivity and nuclear transmutations. He changed the world of physics. My own career in nuclear astrophysics is so dominated by ideas Rutherford started that I cannot dissociate my thoughts of Cambridge from all that he did. For me it was like walking within history.

Only a month after Rutherford first came to Cambridge as a student, Röntgen announced his discovery of x-rays. Rutherford's interests soon turned to the radioactivity that produced them, and in particular to the so-called alpha rays. Rutherford believed that the alpha rays were the primary manifestation of radioactivity, and he was to use them in many brilliant ways throughout his life. After moving to McGill University in Montreal, he began to measure the rates at which uranium and other radioactive elements emit alpha rays. Different radioactive elements have differing activities, in the sense that 1 milligram of radium, say, emits many more alpha rays per hour than does 1 milligram of uranium. In 1903 Rutherford published his first pioneering paper. He and F. Soddy discovered that the activity of each radioelement decreased with time, but in just such a way that the decrease in the activity over a short period of time was proportional to the intensity of the activity! The smaller the intensity, the smaller its rate of decrease was found to be. Rutherford reasoned that in the act of emitting the alpha ray, the radioactive atom changed into another atom that was no longer radioactive. The law of radioactive change was therefore interpreted in the following way: Each radioactive species has a characteristic halflife during which time half of the atoms decay, causing the intensity also to be halved. During the next halflife one-quarter of the original atoms (half those remaining) also decay, leaving the intensity equal to one-quarter of its original amount. The process continues indefinitely until the activity is gone. *But why isn't it all gone?* Why do radioactive atoms still exist? I have their 1903 paper before me now. In part, it says:

> In the processes of radioactivity these different kinds change into one another and into inactive matter, producing corresponding changes in the radioactivity. Thus the decay of radioactivity is to be ascribed to the disappearance of the active matter, and the recovery of radioactivity to its production. When the two processes balance, a condition very nearly fulfilled in the case of the radio-elements in a closed space, the activity remains constant. But here the apparent constancy is merely the expression of the slow rate of change of the radio-element itself. Over sufficiently long periods its radioactivity must also decay according to the law of radioactive change, for otherwise it would be necessary to look upon radioactive change as involving the creation of matter. *In the universe, therefore, the total radioactivity must, according to our present knowl-*

> *edge, be growing less and tending to disappear.* Hence the energy liberated in radioactive processes does not disobey the law of the conservation of energy.

The italics are my own. Rutherford realized there was a very curious problem indeed involved with the fact that natural radioactive atoms still exist. By test Rutherford had shown the activities to be impervious to thermal and chemical effects and regarded them as being immutable properties of the radioactive species, changed only by the decays themselves.

The fact that the intensity decreases in an exponential way (proportional to itself) offers the possibility of a "cosmic clock." By that I mean a clock capable of measuring the time of ancient cosmic events. Rutherford pioneered the techniques for doing this. He had identified the alpha rays as helium atoms (actually *helium nuclei*, but Rutherford had not yet made *that* discovery) by their physical properties of mass and charge and by the curious fact that uranium minerals contain trapped helium gas. In the same 1903 paper he calculated that 1 gram of uranium emits 2,000 alpha particles per second. If this uranium were contained in a large mineral, the alpha rays would stop and be trapped as normal helium atoms. By making that calculation for known helium-bearing uranium minerals, Rutherford was able to conclude that many of these rocks had existed in solid form for hundreds of millions of years. Only in solid form could alpha particles be trapped. Such large ages were surprising to the scientific world and caused much excitement. But that was only the beginning.

The American chemist H. B. Boltwood was quick to follow Rutherford's lead, and in fact they became good friends. Their correspondence is one of the celebrated ones in scientific history. Boltwood was a professor of chemistry at Yale University, where Rutherford gave the Silliman Memorial Lectures in 1905, and was instrumental in causing Yale to offer a professorship to Rutherford. After many considerations, Rutherford declined the offer, to Yale's everlasting loss. Boltwood became certain in 1905 that the other stable product of uranium decay was the heavy metal, lead. Having had the chance to discuss Rutherford's dating technique, he wrote Rutherford on April 18, 1905:

> I have been working a lot over that question as to whether lead is a

> decomposition product of radium and am extremely impressed by the data which I find in support of this hypothesis! . . . If lead can be shown to be a disintegration product of uranium, will it not necessarily follow that all the lead existing on the globe originated in this way. I think that the deductions which can be made from this assumption will make even the metaphysicians dizzy!

At a meeting of the American Chemical Society in 1905, Boltwood gave his findings to the public. Unfortunately, neither Boltwood nor Rutherford were able to carry this argument to its conclusion because they were not able to get a good estimate of the relative abundance of lead and uranium. They continued their collaboration a few years later in Manchester, where Boltwood spent a year after Rutherford moved to a professorship there. Together they struggled to unravel the decay chains of uranium isotopes. An essential stumbling block was that they did not yet have the capability for separating the different isotopes (nuclear masses) of the elements lead and uranium. Today we know that the mass-235 isotope of uranium, ^{235}U, decays to the mass-207 isotope of lead, ^{207}Pb. The halflife is about 700 million years. We also know that today ^{207}Pb is about 10,000 times more abundant than ^{235}U. The ^{235}U is almost gone, but not all gone. Even if all the ^{207}Pb is entirely due to ^{235}U decay, as Boltwood suggested, one only has to go back about thirteen halflives until the ^{235}U would be much more abundant than the ^{207}Pb. One immediately concludes that if the uranium atoms were all created at one time, that time cannot be longer ago than thirteen halflives of ^{235}U, or about 9 billion years. Imagine that! One could have assumed, and many did, that either the uranium atoms have always existed or that they were part of the biblical creation. But by this simple argument we can see that the uranium atoms are no more than about twice as old as the planet earth. Incidentally, the age of the earth itself is now known by a somewhat more complicated variation of this argument involving two different isotopes of lead. It is about 4.6 billion years old, but Rutherford's contemporaries then thought it was much younger.

It was not possible for Rutherford to carry these arguments very far without the mass spectrograph, an instrument for isolating the separate isotopes of an element. This was eventually accomplished after Rutherford moved back to Cambridge, Nobel Prize already in hand, as Cavendish Professor of physics. F. W. Aston, his colleague in the Cavendish, was able, in 1929, to measure the relative abundances of the isotopes of lead in a rare Norwegian uranium mineral. Rutherford was quick to draw the cosmological conclusion. He first concluded that the element uranium has a rare isotope, ^{235}U, that is only 0.28 percent of the abundance of the more prominent ^{238}U. He was also able to infer the halflives of both uranium isotopes. Rutherford still had no explanation of where the radioactive elements came from, but he made a brilliant and plausible suggestion. Since he thought the earth formed from the sun, he supposed that the uranium was created inside the sun. He further reasoned that whatever process created the uranium isotopes would have created them in roughly equal amounts. In an article in *Nature* magazine in 1929, Rutherford concluded:

> It is clear that the uranium isotopes which we observe in the earth must have been forming in the sun at a late period of its history, namely, about 4×10^9 years ago. If the uranium could only be formed under special conditions in the early history of our sun, the actinouranium on account of its shorter average life would have practically disappeared long ago. We may thus conclude, I think with some confidence, that the processes of production of elements like uranium were certainly taking place in the sun 4×10^9 years ago and probably still continue today.

I myself have been very much influenced by these ideas, if only because I have been so much influenced by my graduate school research advisor, William A. Fowler, who was also much influenced by them. Today we feel certain that the earth formed at about the same time and in association with the formation of the sun and that the uranium was created, not by the sun, but by earlier generations of stars which expelled fresh radioactive nuclei like fallout from cosmic thermonuclear explosions. Astronomers see these stellar explosions and call them "supernovas" because they are so much brighter than the more common nova explosions. A supernova explosion may put out as much energy in a few weeks as the sun does in burning for a billion years! The internal temperatures are so hot that the atomic nuclei are transmuted into new atomic nuclei, and thereby the synthesis of new nuclei is affected. Our most terrible terrestrial thermonuclear explosions are infinitesimal in comparison, but even in these we have found that new atomic nuclei are synthesized. Who has not heard of the worry of strontium-90 in cow's milk? It is perhaps less well known that the element californium, which is heavier than the heaviest naturally occurring terrestrial element, uranium, was unknown until the first hydrogen thermonuclear blast at Eniwetok in 1952. A wholly new element had been created by the intense nuclear reactions occurring during the blast. Much more plausible than that fact is the supposition that uranium was synthesized in those intensely more violent heavenly blasts. Rutherford had the right idea; only his scenario was a little wrong.

Rutherford would have been delighted to notice how his methods have been enlarged to include different radioactive species. During the last decade these newer variants have confirmed that the elements were not synthesized at one particular time, but have instead been synthesized continuously since the earliest days of our galaxy. To get a given amount of decay from ^{235}U, for example, one can either suppose that all of the ^{235}U was created about 8 billion years ago, say, or, alternatively, that some of the ^{235}U is older and some younger than that average age. In the latter case one usually assumes that the $^{235}U$was created continuously within the Galaxy —probably by the recurring supernova explosions. The average large spiral galaxy has one every few decades or so. It is hard to pin down the exact age distribution of the ^{235}U nuclei because there are too many possible solutions having the same average age. To constrain the answer requires new information, which is being added to year by year.

The radioactive clocks show that meteorites formed during the same general time that the earth did, about 4.6 billion years ago. These heavenly pieces of rock and iron revolve in orbits about the sun that sometimes allow them to fall into the earth's atmosphere. At night they look like falling stars, and are often called just that. I well recall the night of August 11, 1972, when Annette counted forty-two of them in a single hour when we were on holiday with her parents at Blaavand on the west coast of Denmark. Most of them were pea-sized objects that vaporize in a few seconds of flaming heat as they scrape through the earth's atmosphere. Occasionally a large one falls to the ground intact, is discovered, and generally winds up in one of our scientific museums. A subset of these, the type-1 carbonaceous chondrites, named Alais, Ivuna, and Orgueil, provide our best indication of the relative abundances of the elements in the solar system. These meteorites have been subjected to so little high-temperature chemistry that the relative abundance of the elements within them remain undisturbed from their primitive values, unlike the surface of the earth, which has seen one element fractionated from another by chemical forces in the tortured evolution of the earth. Study of the radioactive isotopes in these meteorites confirms their age, and even more importantly, provides evidence of the continuous synthesis of new elements.

Professor John Reynolds and his coworkers at the University of California at Berkeley have led the way through the maze of this astonishing evidence. It began when they found that the xenon gas trapped in the meteorites did not have the same isotopic composition as xenon in the earth's atmosphere. Very little xenon gas was trapped in the meteorite during its formation as a solid object, so it was quite noticeable that the isotope ^{129}Xe was anomalously overabundant in some. By careful experiment they showed that this ^{129}Xe is due to the decay of a radioactive isotope of iodine, ^{129}I. Because its halflife is only 17 million years, less than 1 percent of the total age of the earth and meteorites, no detectable quantities of ^{129}I remain today. However, the ^{129}Xe that it decayed to *is* detectable, and its presence there suggests that the meteorites formed with substantial amounts of radioactive ^{129}I in their iodine content. If the meteorites themselves did not contain actual radioactive iodine, they must alternatively have formed from preexisting grains that long ago contained the radioactive iodine and that still contained the daughter ^{129}Xe. About one iodine nucleus in 10,000 was the mass-129 isotope when the meteorites formed. This may not sound like much, but it is far more than would be expected if all the elements had been synthesized at one time, say 8 billion years ago. The basic Rutherford question of "Why aren't they all gone?" rears its head again, and the most satisfactory answer is that some of the radioactive ^{129}I was synthesized a mere few hundred million years before the meteorites formed. Stated more meaningfully, some of the iodine nuclei on earth are no more than a few percent older than the earth itself!

Similar suspicions were confirmed when other anomalies in the xenon isotopes were discovered. It appeared that some heavy radioactive nucleus, heavier than uranium, was also chemically trapped in the meteorite's solidification. Afterwards the nucleus split in two by an act of spontaneous fission, which, though sounding quite exotic, has been thoroughly studied in nuclear physics laboratories. Such nuclei are simply too massive for stability, and they break apart into two medium-heavy fragments. A fraction of these fragments are isotopes of xenon, a feature which again allows them to be detected in meteorites simply because the meteorite contains almost no trapped xenon gas from other sources (natural xenon). But the path to discovery was tortuous, and required long years of expensive and delicate experiments. Natural xenon has seven different isotopes, which have the same nuclear charge and chemical properties but differing nuclear masses. By a very careful mass spectrometric analysis resembling the technique, spirit, and importance of Aston's pioneering work in the Cavendish, it was determined that the fissioning nucleus had produced in the meteorites only the four most massive xenon isotopes. The relative yields had been in the same order as the increasing nuclear mass: ^{131}Xe, ^{132}Xe, ^{134}Xe, and ^{136}Xe. Just as Rutherford had inferred the existence of ^{235}U from the lead isotopes, many modern Rutherfords inferred that the extinct fissioning nucleus in the meteorites had been an isotope of plutonium, ^{244}Pu. One of the main reasons is that its halflife, 82 million years, is easily long enough that the meteorites will have had time to form while a substantial concentration of it still existed. Nevertheless, the identification was in doubt until May 1971 when the Reynolds group published the results of a difficult and sensitive experiment. They were able to get a 13 milligram example of ^{244}Pu, which had never been discovered naturally on earth but which could be made from uranium in nuclear reactors. Within it they could measure the relative yields of xenon isotopes. They found them to be exactly the relative abundances of the anomalous xenon in the meteorites. That solved the puzzle. Armed with that assurance, another group of scientists was able to find natural ^{244}Pu in the earth. Because the solar system formed sixty halflives ago, the average concentration of ^{244}Pu is expected to be only $(½)^{60}$ of its concentration when the meteorites formed. That average concentration is too small to find, being only about 40 grams of ^{244}Pu in the entire earth! Amazingly enough, however, they found some in a mineral called bastnasite, that has, in the long and complicated chemical history of the earth, concentrated the plutonium. It was a great detective story, and Rutherford would have loved it.

The cosmological significance of these discoveries is both simple and profound. When the primeval meteoritic abundance of ^{244}Pu is considered in conjunction with that of ^{129}I, species now effectively extinct, it can be concluded that about 100 million years elapsed from a time when a portion of the interstellar gas of the Galaxy collapsed into a gas cloud that was to form the solar system and the time when the newly formed meteorites were cool enough to retain noble xenon gas. This well-determined interval for the formation and cooling of the meteorites poses a firm challenge to the theories of the formation of solid objects. The only escape from this conclusion would seem to be if some small grains rich in xenon anomalies existed before the solar system and were incorporated into the meteorites when they formed. In that case, the meteorites must contain small cooled chunks of ancient stellar explosions! I described this exciting possibility in an invited address to the Anaheim meeting of the American Physical Society in January 1975. It is, I think, one of my own most creative ideas, because it gives a new interpretation to the meaning of the extinct radioactivities and to modern observations of interstellar dust grains. Curiously, however, it is encountering strong *emotional* resistance because it does, if correct, invalidate earlier interpretations. Scientific objectivity will eventually converge on the true picture.

The implications of cosmic radioactivity do not stop even with the origin of the solar system. Because ^{244}Pu, ^{238}U, ^{235}U and ^{232}Th, the last a long-lived isotope of thorium, are all synthesized in comparable amounts in the supernova explosions, their relative abundances when the solar system formed reveal their production history. There had to have been some production shortly before the solar system formed or there would be insufficient short-lived ^{244}Pu, and there had to be substantial production 10 billion years ago to allow enough of the ^{238}U to decay, so that it can today be much less abundant than ^{232}Th, as observed. When all of the information is assembled and analyzed in full mathematical detail, one reaches these conclusions: The atoms of matter as we know it did not always exist, nor were they created in a single event at any time. They came into being over an extended period of time beginning between 10 and 15 billion years ago (two to three times the age of the earth) and continuing with certainty up until the time the solar system was formed and probably until today. This age pattern for the elements must be taken as the age pattern for the Galaxy as we know it because it reveals the history of exploding stars in the Galaxy. It confirms other evidence that our Galaxy is between 10 and 15 billion years old. If other galaxies have the same age, then this age must be something like the age of the universe, if, in fact, such a concept has meaning at all.

In my graduate school years, Professor Fowler, to whom I shall return later, introduced me to the mysteries of the abundances of the heavy elements. He had spent a sabbatical year in Cambridge, under the spell of the influence of Rutherford. His curiosity about the age of the elements infected me. In my office in Kellogg Lab, poring over the abundance charts one day in 1962, I noticed that one isotope of the heavy toxic metal osmium was too abundant to be accounted for by a theory of the origin of the elements that Professor Fowler and I were working on. It immediately dawned on me that the reason was radioactive rhenium, another heavy metal, which decays to osmium with a halflife greater than the age of the universe. The elements have existed just long enough for this decay to make that one isotope of osmium overabundant. Somewhat stunned, I realized I had just discovered a new radioactive clock. It was a very lucky moment in my life. It still excites me today as nuclear physicists continue calibrating the clock by measuring the needed properties of the osmium nuclei. The most important step is that of measuring the relative probabilities with which the two different osmium nuclei captue free neutrons. This measurement requires pure samples of the two isotopes, a difficult challenge for isotope-separation technology. Only now, in 1974, fully ten years after I described the method of this nuclear clock, have separated 2-gram samples of these osmium isotopes been prepared. Soon, in nuclear laboratories at Livermore and at Oak Ridge, nuclear scientists will be measuring the age of the elements with these samples. What an imaginative opportunity—to measure the universe in a piece of metal!

When one leaves the physics library of the Cavendish Laboratory the majestic full portrait of Rutherford by Oscar Birley bids one good-bye. He stands in a Cambridge gown before his apparatus, one

Portrait of Lord Rutherford of Nelson, O.M., F.R.S., Cavendish Professor 1919–1937, painted by Oswald Birley. (U.K. Atomic Energy Authority)

hand on a book. It reminds one that a lot of physics happened right there. An important step in cosmology happened there too, when mankind faced that simple but insistent question: "Why aren't they all gone?"

CHAPTER VII
THE INVENTOR OF CONCEPTUAL EQUALITIES

The incomparable Irish poet, William Butler Yeats, wrote in a letter to Olivia Shakespear: "I am still of the opinion that only two topics can be of the least interest to a serious and studious mind—sex and the dead." There is much in what he says, and Yeats's own impassioned poems confirm that, for him at least, it was the truth. A romantic physicist, on the other hand, might opt instead for love, space-time, and equalities. Love provides the fulfillment, the inner security, the contentment, and the sense of belonging to nature so characteristic of those that experience it and so yearned for by those less fortunate. Love extends beyond the personal and familial and embraces all of existence. To one bearing love, the commonplace things of nature take on an exquisite fascination and joy. Each of the few great physicists and cosmologists I have known have it. One could with much truth call the joy childlike, although in these cases one is aware of an almost devout respect for natural phenomena that is a part of their great intellects and advanced physical knowledge. One doubts if Isaac Newton found much interpersonal love in life, but its other manifestations show clearly in his own words:

> I do not know what I may appear to the world; but to myself I seem to have been only like a boy playing on the sea-shore, and diverting myself in now and then finding a smoother pebble or a prettier shell than ordinary whilst the great ocean of truth lay all undiscovered before me.

Space-time is the intellectual structure used to describe events—it happened at *that* point in space at *that* time according, for example, to the surveyors of the Royal Admiralty and to Her Majesty's Royal Timekeeper. Someone who records when and where a thing happens locates an event in space-time. Scientific experiments are elaborate planned forms of such observations. Newton himself eloquently imagined space as full of three-dimensional graph paper, so that the distance between two events could be measured with unambiguous assurance; moreover, absolute time was regarded as passing equitably and unambiguously at all points in space. In an astonishing sequence of virtuoso insights, perhaps the most artistically creative the world has seen, Albert Einstein changed all of that. Space-time itself remains, for men always want to locate events, but its absolute structure was shown to be a house of straw.

Equalities are the ways of making quantitative or qualitative statements, and, as such, they lie at the heart of the scientific adventure too. They enable mathematics to be attached to measurements in space-time. For example, "It's hotter here than it was there," might become the equality in space-time, "The temperature in Cairo at noon today equals the temperature in Athens at 1:00 P.M. yesterday plus 7° centigrade." It's a mouthful, and perhaps not worth the effort in this case, but it obviously says a lot more. Even more importantly, equalities symbolically express an idea, a relationship between quantities whatever their values. Thus Newton's second law, $F=ma$, expresses the relationship between the acceleration of a body having mass m and the force necessary to produce that acceleration. When an equality is between concepts themselves, moreover, thought is sometimes revolutionized. Such rarities are not given to the lives of most physicists, but Einstein was no ordinary physicist. He did it often! He *invented* conceptual equalities that revolutionized the human view of the universe, and because of that, four of his equalities are never far from my mind. I embrace them daily with mixed belief and astonishment because, amazing as this may sound to one who has not experienced it, they enable me to regard everything with increased wonder and affection. Yes, *affection*.

When a new concept provides understanding of many of nature's wonders, it does bring increased inner security; for then one has seen more order in things, and one can regard them as affectionate extensions of his consciousness. Einstein himself found great joy in his work, even though its high ambition confronted him with countless frustrations. The experience of unifying and increasing knowledge is an almost religious one. It clearly was so to Einstein, for he is reputed to have frequently referred to God in comments upon scientific matters, although his meaning for that concept was rather abstract and unspecific. Often it seemed to stand for a criterion of logical simplicity. When finally asked for clarification he wrote:

> It seems to me that the idea of a personal God is an anthropological concept which I cannot take seriously. I feel also not able to imagine some will or goal outside the human sphere. My views are near those of Spinoza: admiration for the beauty of and belief in the logical simplicity of the order and harmony which we can grasp humbly and only imperfectly. I believe that we have to content ourselves with our imperfect knowledge and understanding and treat values and moral obligations as a purely human problem—the most important of all human problems.

It was, in any case, this criterion of logical simplicity that motivated Einstein to his revolutionary ideas.

For seven years, between 1902 and 1909, Einstein lived in Bern, Switzerland and worked as a Technical Expert Third Class in the Swiss Patent Office, later promoted to Technical Expert Second Class. I can hardly write that down without laughing. Yet inappropriate as it may seem, it may have been best for Einstein that way. Repelled by the pressures of academic competitiveness and routine learning, alienated by the overemphasized military mentality in the German state, Einstein gave up his German citizenship while still a teenage boy and became a Swiss citizen. There, encumbered only by the routine tasks of the Patent Office, his mind was free to concentrate on the puzzling equalities that perpetually tantalized him. I must admit some envy, in fact, because I sometimes find that my own life as a professor is sometimes too full of various pressures to allow much time for free thought. Einstein recognized this, and later spoke nostalgically of the Patent Office as "that secular cloister where I hatched my most beautiful ideas." Beautiful they were, but when Einstein submitted them to Bern University as an inaugural thesis toward beginning a university career there, they were rejected as incomprehensible.

The incomprehensibility of Einstein's ideas arose in large part from his startling conclusions concerning the nature of time. It had,

Einstein in the Bern Patent Office

since Newton, been thought that time is an absolute measure of when events happen—absolute in the sense that if one event is seen to precede another, it definitely and unambiguously *does* precede the other. Yet Einstein showed convincingly that if two events occur rather far apart and rather close together in time, the order of the events can appear reversed to different observers. Since the two observers had been argued to be equals by Einstein, he concluded that it is meaningless to claim absolutely that two events happen simultaneously.

Einstein arrived in this seemingly awkward position by having the physical intuition to invent two new equalities in human thought and to courageously stick by them because of their logical simplicity. The first was his instinctive assertion that it is *impossible* for a moving laboratory to tell in any way whether it is at rest or moving smoothly along at constant speed. We all now realize it would be no problem to play catch in a Boeing 747. This fact had been known to Newton insofar as the motion of massive bodies was concerned, but when Michelson and Morley failed to detect the effect of the motion of the earth on the apparent speed of light rays, Einstein quickly took it to be true for any measurements whatsoever, including those with light. Thus, if two observers move toward each other with high speed, they can in no way tell if the first is stationary and the second moving, or if it's the other way round, or if they both move. This means that the descriptions they give to events and to the laws of physics are equally valid. They are equals. In 1905 Einstein called this *the principle of relativity.* It seems simple enough, or at least it will until we realize the consequences.

Einstein's second great equality is that the speed of light always equals the same value *c*, regardless of the relative velocity of the source of the light and of the observer. Later, in constructing his general theory of relativity, Einstein was to drop this assumed constancy for observers who are accelerating or who watch light move through gravitational fields. But, strict constancy for the speed of light was consistent with the Michelson–Morley experiment, for which others were attempting to construct elaborate explanations. Einstein's audacity was to propose that there was nothing to explain; it's just a fact, and in itself, the simplest and most logical of all possibilities. Although he is not known to have said so in this case, he probably thought that God would do it that way because it's simpler. But on second thought we see that something new lurks in this postulate because we know, or at least *we think we know*, that a ball thrown 100 miles per hour toward the front of a Boeing 747, which has an air speed of 600 mph, itself moves at 700 mph through the outside air. If this is to not be true for light, something must give; indeed, all the old notions of lengths and times come tumbling down.

It is not difficult to see how these two plausible assumptions of Einstein attack the old concept of absolute time. Suppose, for example, that inhabitants of Earth and Mars wish to synchronize their clocks by timing bursts of x-rays from solar flares on the surface of the sun. From the radii of the orbits of Earth and Mars about the sun, they both know that the x-rays reach the Earth 8 minutes after the solar burst and that they reach Mars, whose orbit lies outside the Earth's, 12 minutes after the burst. By agreeing that the Earth inhabitants shall start their clocks from zero when the burst arrives and that the Mars inhabitants shall start theirs from 4 minutes when the burst reaches them, they all have satisfactorily synchronized their clocks. From that time on the inhabitants of both planets can regard events that are simultaneous to them both; for example, they could agree to simultaneously flash a strong laser beam toward some other object.

Suppose, however, that the entire solar system moves with high speed relative to an astronomer in some other system. For simplicity of thought, suppose Mars lies in its orbit on the opposite side of the sun from the earth, so that all three bodies lie on a straight line with the sun intermediate, and suppose also that the entire solar system speeds in the direction of that same line from Mars toward Earth. The astronomer in another system also sees the x-ray burst of the sun, and, in accord with Einstein's second proposition, he sees the x-rays move outward with constant speed such that the front of the x-ray burst forms a spherical shell *about the place where the sun was when the burst occurred.* He quickly sees that the Earth inhabitants and Mars inhabitants make an apparent mistake; the x-rays travel different distances than they thought. In the 8 "Earth minutes" required for the burst to reach the Earth, the outside astronomer sees that due to the high speed of the entire solar system the Earth has moved farther away from the place where the sun was when the burst was emitted, whereas Mars has moved closer to that same origin of the burst. He sees clearly that the distances traveled by the x-rays do not stand in the same ratio as the radii of the orbits of Earth and Mars. Since the distance traveled at the speed of light is just proportional to the time of travel, the outside astronomer sees that the Earth inhabitants and the Mars inhabitants have incorrectly synchronized their clocks. The subsequent events that they believe to be simultaneous in fact occur at considerably different times. Everyone has acted quite sensibly, however, so who is right? According to Einstein's principle of relativity, the observers are equals, and, in reality, they cannot even tell if the solar system whizzes past the outside astronomer, or if the outside astronomer has whizzed past the solar system in the opposite direction. Thus both must be right! No wonder that Bern University returned the thesis as incomprehensible.

Einstein remained calm in the face of what he described as "only apparently irreconcilable" principles. He found his two principles to be more plausible than the concept of an absolute order of events. If time is nothing more than the measured duration of the light ray's motion between the locations of two events, then there is no paradox. One then regards the order of events as a relative quantity, rather than an absolute one. Events on Earth and Mars separated by, say, one minute according to the agreeing solar system inhabitants may appear to be separated by different times and, indeed, even to occur in reverse order to an outside astronomer in fast relative motion. I have continually had to remind myself to not think of this as an optical illusion, for "illusion" implies that there is a "correct" sequence of events that is being distorted. We quickly appreciate, however, that the revolution of relativity is much more than carnival magic; if Einstein's postulates are correct, and every evidence known to science says that they are, there can in principle be no absolute statement that two events at different places happen at the same time.

When I first encountered this idea, I was a student at Southern Methodist University. At the time I could not accept it because I didn't actually understand it. In particular, I was quite certain that when I shot the moving metal duck at the carnival rifle range, the event of my squeezing the trigger definitely proceeded that of the bullet knocking down the duck, and I could not accept any theory that held it otherwise. In time I learned that Einstein's theory makes no such bizarre demands. Cause and effect are not undermined because no causitive influence travels faster than light. The order of two events can be relative only if they are separated by sufficiently great distance and/or sufficiently short time so that a light ray cannot make it from the first event to the place of the second before it happens. The order of the events on Earth and Mars separated by one Earth-minute is relative, because a light ray cannot make it from Earth to Mars in one minute. On the other hand, the light ray could travel from the squeezing trigger to the duck much faster than the bullet can, so it turns out that every outside astronomer would see the gun fire before seeing the duck fall. In this case only the measured duration of the bullet's travel is relative; to me squeezing the trigger it seems $^{1}/_{100}$th of a second, but to a speeding observer it might seem to take 10 seconds. That still seems pretty fantastic, however, so you can imagine my interest when in later years I learned that this effect has been confirmed by direct measurement. The two events can be the production and decay of the μ-meson in a modern accelerator laboratory. By measurement the μ-meson at rest in the laboratory lives only two millionths of a second after production before it decays. That's a very short time by human standards, but easy to measure with modern precision electronics. When the μ-meson is traveling at speeds almost as fast as light in the accelerator, however, the situation is quite different. By the principle of relativity this situation cannot be distinguished from that of the laboratory rushing past the μ-meson at almost the speed of light, and the apparent lifetime of the μ-meson should be much longer. It is. In the accelerator μ-mesons live as long as several thousandths of a second, fully a thousand times longer than they appear to live at rest. The exact numerical answer is precisely that predicted by relativity theory. The same effect is observed when cosmic ray μ-mesons penetrate the earth's atmosphere. The faster moving ones appear to live longer than the slower moving ones. More recently, to calm remaining skeptics, a precision atomic clock has been carried in a jet aircraft and found to advance more slowly than its identical twin at rest, exactly as predicted. What a triumph for human thought!

Since simultaneity and the duration between two events are relative concepts, length must be also. There is no other way in which light can present the same velocity to all observers. Suppose the Earth inhabitants and Mars inhabitants decide to measure the distance between Earth and Mars when they are in any random relative positions in their orbits. Since they have synchronized clocks they do this by having the Mars inhabitants emit a laser pulse toward Earth at a specified time when the Earth inhabitants start a stop-watch that will stop when the pulse of light arrives. Knowing exactly the time the light took to make the journey from Mars to Earth, they will then know the distance accurately. But alas, the outside astronomer can only shake his head in pity because he has already noted that the clocks were incorrectly synchronized. The Solar System inhabitants seem to him to get the wrong distance because they timed the interval of transit incorrectly. The theory of relativity presents no contradiction in this, for again both are correct—length is only relative. Even the length of a steel rod has no absolute meaning. There is no stopping the flood of relativistic quantities, for everything is measured with lengths and times. Apparent forces, accelerations, and even masses are all relative. Amazingly, the laws of physics themselves are less fickle. All equal observers find the same equations for the same laws of physics. The same equalities work for each observer, even though the values of the individual quantities in the equations are relative. Thus, when the Solar System inhabitants make their midcourse correction to the Pioneer spacecraft to Jupiter, they use Newton's second law, $F=ma$; on the other hand, the external astronomer uses the same equality but with different values for the thrust F, the spacecraft mass m, and its acceleration a. This is the "special theory of relativity."

Later in the same year, 1905, Einstein published a calculation showing that if an atom emitted light having energy E, its mass M must decrease by the amount E/c^2, where c again represents the velocity of light. One can use the relativity expressions for momentum and energy to establish this equality. This earth-shaking conclusion is now rediscovered yearly from such a calculation by university students around the world. Within two years Einstein had realized that having two kinds of mass was redundant and unesthetic, and he argued that not only did energy of any kind have the effect of mass but also that all mass, even the most common handful of sand, had energy according to the famous equality $E=Mc^2$. The very concepts of mass and energy are thereby made equivalent, since the square of the velocity of light is only a constant; but the energy contained in mass is, by society's standards, enormous. A grain of sand is a reservoir of mass energy equal to the chemical energy of 10 tons of gasoline!

I have often wondered how many physicists at that time really believed the sweeping extent of this equality when Einstein quietly dropped it on the world. Rutherford in Canada was already detecting the very phenomenon, alpha decay, that Einstein guessed could liberate sufficient energy to make a test of his equality possible. Yet it was to be almost three decades before Rutherford's colleague Aston could measure the masses accurately enough to confirm this prediction. They would find, for example, that when the mass-238 isotope of uranium decays to the mass-234 isotope of thorium by the emission of the mass-4 alpha particle ($^{238}U \rightarrow {}^{234}Th + {}^{4}He$), some of the "rest mass" is converted into energy of motion of the alpha particle. The approximately 238 atomic mass units of the ^{238}U actually weigh somewhat more than the approximately 238 atomic mass units of the $^{234}Th + {}^{4}He$. When this disappearing rest mass is multiplied by c^2, one found an energy exactly equal to that of the motion of the ejected alpha particle. If one could *weigh the motion* of the alpha particle, one would find that it carries exactly the lost weight. This bizarre concept has been confirmed countless times in nuclear physics laboratories, where one finds that the high speeds of particles inside the nucleus do contribute to its weight. A more dramatic demonstration in today's laboratories is particle production. In these carefully measured events, an energy of motion is converted into the rest masses of newly created particles—always in accord with the Einstein equality.

I wonder if you share my interest in finding that this esoteric concept is much more than a mental toy for the world's physicists. It lies at the heart of life and simultaneously poses the threat of man to mankind. The friendly sun, whose warming rays stimulated and sustain life, would long ago have burned out were it not converting mc^2 to E. Five million tons of solar mass disappear each second and are replaced by light energy radiating outward from the sun. Nuclear power plants on earth attempt the process in miniature to help run our air conditioners. At Hiroshima and Nagasaki the horrible military possibilities were initiated and now more powerful bombs wait in threatening stockpiles. Only a fraction of an ounce of matter need disappear to devastate a city. Certainly Einstein never meant the bomb to be used when he used his influence in a letter to President Roosevelt, for Einstein was himself a devout pacifist. Without doubt he was in great conflict; he had fled the Nazis in 1933 and regarded them as an evil worse than war. In 1933 for the second time in his life Einstein had renounced his German citizenship, which he had previously reestablished to take the Berlin professorship. Within this setting he urgently warned President Roosevelt in 1939 of the possibility of a nuclear bomb; but, when it was later used against the Japanese, he despaired. It would not be correct to think, however, that Einstein's equation, deduced from the logical principles of relativity 40 years before Hiroshima, made the bomb possible. The laboratory discovery of the enormous energy released by nuclear fission would itself have sufficed, given man's combative instincts. Nonetheless, Einstein expended considerable effort in his late years promoting awareness of the dangers of nuclear armament.

The foregoing thoughts pertain to Einstein's "special theory of relativity." The enlargement to the "general theory of relativity" is not an easy one for us to make. The attempt may be aided by a childhood episode that strengthened my own intuition in this regard. The carnival rides along the midway at the State Fair of Texas provided many a boyhood thrill. Some tantalized the curiosity as well as the stomach. Let me reminisce about one, of whose relevance you may be the judge. This heart-pounding experience was a giant vertical cylinder that could be set in rotation. People stood along the vertical walls of the cylinder as it began to rotate, faster and faster, until each one was pinned against the wall. Ever faster it rotated until that dramatic moment when the floor was lowered away. There we all stuck like flies to the walls, laughing hysterically. My brother and I carried wadded balls of paper that we hopelessly tried to throw to each other inside that rotating cylinder before it reached full speed. It was maddening. So curved was its apparent path that even a rapid toss was all but impossible to catch. On the way home we invented a comic dream that lives to this day in my mind—a baseball game played on a giant rotating turntable. We decided that second base would be a good spot for the center of the turntable. The extra curvature to the pitched ball would challenge both pitcher and batter. But it was the play action that really caught my fancy. Ground balls and the throw to first base would be exciting enough, but imagine a high pop fly to short left field!

By playing with several different types of balls, a pitcher on this field could notice a curious thing. Any ball, be it a baseball, golf ball, or tennis ball, would, when pitched with the same speed, suffer the same curvature of its path toward home plate. You have to think about that one a bit. Suppose the field is the entire Astrodome so that the players, who can be imagined to have lived their whole life in the stadium, do not know that it rotates. What would they think about the peculiar trajectories of thrown balls, assuming that, unlike real baseball players, they have been educated in the laws of physics? They would learn Newton's law $F=ma$, and they would observe unusual fields of force in their stadium. A ball sitting at rest on the field would begin to roll farther away from second base, and a force would be required to hold it in place. They might name it "centrifugal force," and they would find it to grow stronger at greater distances from second base. The observed curvature of the path of thrown balls would require an even more complicated force field; it would depend on both the speed and direction of moving balls. They might name it "Coriolis force." Players might play for many seasons and develop a keen instinct for coping with these forces.

Through all of this imagined scenario, a young pitcher named Lefty Einstein might have spent his time in the dugout puzzling over the fact that these forces produce the same acceleration on different objects. He knows that all balls pitched with the same speed toward home plate curve by the same amount. He knows that if he throws a golf ball and a baseball straight up to the same height, they rise and fall on identical curved trajectories, hitting the ground at exactly the same spot some distance away. Those facts give him no peace. He puzzles because Newton's law $F=ma$ requires that, since each ball is observed to move with the same acceleration a, the force field must be producing a force on each ball that is proportional exactly to its inertial mass m. The inertial mass measures an object's *resistance* to acceleration. Thus the puzzling feature to Lefty would have been that the force on each object is directly proportional to its intrinsic resistance to being accelerated. He would have regarded this as a very complicated explanation of a basically simple fact. The trajectory of the ball might have seemed to him more fundamental than the forces and masses used to explain it. For several seasons he might have pondered this situation.

Finally it comes to him. The motion of the balls is exactly what a rotating stadium would produce. He points out to his incredulous team that a simpler description of nature than the one they have been using is possible. There exists, he claims, an "inertial frame of reference" in which balls move on straight lines, and their paths only appear curved with reference to the stadium floor, which is not an "inertial frame." Thus, the inertial frame of reference became a concept of great importance in physics. An observer in such a frame will see that two objects released at rest a small distance apart do not accelerate away from the observer or away from each other. In the rotating stadium, for example, this is not the case, as two baseballs placed on the field immediately begin to roll away from each other. The centrifugal and Coriolis forces are reduced to fictions that appear because of the description in a rotating reference frame. The theory explains in a simple and natual way why balls move on the same trajectory, because, in the nonrotating inertial frame, they all move on the same straight line. It becomes known around the stadium as the "rotational theory of relativity" and is much admired by those players that understand it. Stadium philosophers would also have been greatly attracted to arguments about whether the stadium "really rotates" or if the observed paths of motion are only coincidentally the same as those generated by rotation. If they could never leave their Astrodome, they could never decide.

This parable may or may not be enlightening and amusing since it was determined by a personal childhood fantasy. By my own experience this fantasy was more help in understanding Einstein's "general theory of relativity" than many a learned treatise. The real Einstein theory includes not only rotation, but, much more importantly, gravity as well. Einstein knew the parable of rotation, and he also realized that gravity has the same puzzling feature. Golf balls and baseballs thrown with the same speed and direction from the top of the tower of Pisa follow identical trajectories. These trajectories were commonly analyzed in terms of Newton's laws of motion, wherein it was assumed, just as Newton had ingeniously outlined three centuries earlier, that the force of gravity on a body was for some unknown reason proportional to its resistance to being accelerated. Then the law $F=ma$ gives equal gravitational acceleration a to all bodies so that, miraculously, all follow the same trajectories. This was the same peculiar property of force fields that could be eliminated in the rotational problem by discovering the inertial reference frame, and Einstein conjectured that the force of gravity might be just as much a fictitious property of having chosen the wrong reference frame as the centrifugal force had been. In 1907, while still in the Bern Patent Office, he made the idea clear to others with his famous parable of the rocket ship (or *elevator*). If a rocket ship accelerates in free space, all objects floating within fall toward the floor with equal acceleration a, which is nothing more than the acceleration of the rocket past the objects within. All astronauts, young and old, today regard this fact as commonplace. On reaching the floor, or a table top, they are pinned against it by the acceleration, much as I was pinned against the wall of the rotating cylinder. The acceleration gives everything an apparent weight. This situation is apparently identical to that of the rocket ship standing on the earth before its launching, when all objects fall toward the floor with equal acceleration a. Only in this case, the acceleration is attributed to gravity. By studying falling objects, the astronaut within cannot tell if they are falling because of gravity, or because the rocket is accelerating in space, or both.

Here Einstein made a great leap towards another conceptual equality. He proposed that a uniform gravitational field is equivalent in *every* respect to an accelerated reference frame; that is, that the astronaut can in no way whatsoever (except by looking outside) decide if his rocket laboratory sits on the earth or is accelerating in space. No experiment he can perform, not only with falling balls but also with magnetic fields, superconductivity, light rays or whatever, can distinguish gravity from acceleration. Einstein called this bold assumption the "principle of equivalence," and his faith in its logical simplicity was his guiding light in constructing the general theory of relativity. It was, however, a much more difficult problem than that of measurements of length and time by moving observers, and it took him many years to find a satisfactory formulation. He constructed a theory in which gravity, like the Coriolis force, is not a genuine force but only one that seems to exist because the observer is not in an inertial reference frame. The Apollo astronauts coasting toward the moon have found their inertial frame, and they experience locally no gravity at all.

The theory is quite beautiful, but difficult to visualize because of its necessarily mathematical structure. The curved trajectories of objects are accounted for by a geometrical structure within space-time itself. They fall along the shortest path between two points, but the shortest path is not straight because large masses cause space-time itself to be curved, with the curvature greater nearer to the mass. Because space itself is curved, even light rays travel along curved paths. The necessity of this is clear from Einstein's principle of equivalence because if the astronauts shine a beam of light across their accelerating spacecraft, its path must appear curved by the acceleration of the spacecraft—and so it must be with gravity too. The theory accounts for every success of Newton's theory, it predicts many new phenomena that appear to be correct, and it explains simply why different falling objects fall along the same trajectory.

Perhaps no more dramatic moment has ever occurred in pure science than in 1919, when astronomers first tested the theory's unusual prediction that light from a distant star behind the sun would be bent as it moves past the sun. Since one cannot normally see stars behind the intensely bright sun, it was necessary to utilize a total eclipse of the sun when the moon moved in front of it. A photograph of stars behind the darkened sun was then compared with their previously recorded positions when the sun was not there—and indeed, Einstein's new and revolutionary theory appeared to give the correct answer, and Newton's the wrong answer. It made the newspapers and caught the public's fancy as few events in scientific history have. Almost overnight Einstein was an idol and a hero to the man on the streets. It was the first time when Newton's theory was widely regarded as having been proved wrong, although in fact the general theory of relativity had already scored a great success with the motion of the planet Mercury. It explained that Mercury's elliptic orbit should not exactly repeat itself, as the Newtonian theory held, but the most distant point in its elliptic orbit should itself revolve very very slowly around the sun. That Mercury actually did so had been known for a very long time, but the situation was not originally regarded as conclusive because of the large number of big corrections that need be made to Mercury's orbit due to the gravitational influence of the other planets. Indeed, I must report that today *neither* observation is regarded as conclusively proving Einstein's theory to be correct, although they do show convincingly that Newton's theory is wrong. Repeated measurements of the bending of light, for example, give varying results near Einstein's value. Both observations are exceedingly difficult, and the numerical accuracy of the results is perhaps not great enough to favor Einstein's theory over some other non-Newtonian theory. Meanwhile, the theory scores other triumphs, so we must hasten on.

One dramatic result, as Einstein showed from the principle of equivalence, is that time passes more slowly as the gravitation becomes stronger. Imagine that in the accelerated rocket laboratory a regularly pulsing source sends flashes of light from the floor of the rocket toward its ceiling. Because of the acceleration of the rocket, each pulse must travel a greater distance (viewed externally) than the preceding one. This happens because the ceiling accelerates to a slightly higher speed to receive each pulse than the floor had when it was emitted. Pulses emitted one second apart on the floor will be received at intervals greater than one second on the ceiling. By the principle of equivalence the same thing would happen if the laboratory stands in a gravitational field. The easiest way to check this has been with light emitted from atoms, where the frequency of the wave replaces the pulsing source. Careful experiments have shown that light emitted with known frequency on the floor is received at a lower frequency at the ceiling. This phenomenon is called the "gravitational redshift"—redshift because lowering the frequency of visible light makes it more red. It does not mean that the light actually *is* red—only that its frequency is decreased. Scientific terminology is, like common speech, often based more on historical mystique than on logic. In actual fact, the "light" used in this experiment was quite invisible gamma rays from the atomic nucleus. The phenomenon has been studied in visible light from the surfaces of stars. The white dwarf star Eridani *B* has a mass about half that of the sun's, but a radius only a few percent of the sun's. The gravity at its surface is about 500 times greater than on the surface of the sun, and the gravitational redshift lowers the frequency of light from each atom by an easily measurable 0.7 percent. When an atomic clock was recently flown in a jet aircraft, it confirmed not only that the air speed lowers the apparent rate of the clock, but also that the clock ran more rapidly due to its altitude. Strange goings on! But this amazing intellectual achievement comes more from the principle of equivalence itself than from the full general theory of relativity. Einstein himself predicted this redshift before he developed the general theory. There are now other theories that are also consistent with the principle of equivalence, so the most one can claim on the basis of the gravitational redshift is that Einstein's theory is in agreement with it. So important is this principle of equivalence that it has been tested with painstaking accuracy. Largely due to Dicke and his colleagues at Princeton University, it is now known that the accelerations of two different balls falling due to gravity do not differ from each other by more than one part in 100,000 million!— so probably they are exactly equal. This accuracy is already sufficiently great to prove that the energy of motion of subatomic particles has weight equal to its inertia against acceleration. Einstein's rocket ship wins again!

In just the last few years another phenomenon depending upon more details than those provided by the principle of equivalence has come under study. Because of the curvature in space-time, not even the velocity of light is a constant in Einstein's general theory. What a heresy from the man who earlier revolutionized physics by assuming the *constancy* of the speed of light! But that was for observers without acceleration or gravity. For arbitrarily moving observers, even the speed of light becomes variable. The deflection of light by the sun is not primarily due to the curvature of space, but also because the waves propagate more slowly nearer the sun. With the development of the almost unbelievably sensitive techniques of interplanetary radar, it was realized that this variable speed suggests a new test of the theory. Radar echoes from the planets can be accurately timed. When the echoing planet passes on the other side of the sun, the reduced speed of light should cause the echo to be received later than the Newtonian expectation. The idea works, and the results provide the strongest evidence yet that in the general theory of relativity lies the correct theory of gravity.

The redshift of the light from the surface of a white dwarf star suggests the most bizarre prediction of the general theory of relativity—the "black hole." The world of astronomy is now vigorously searching the skies for physical evidence of these collapsed objects. The redshift effect in dwarf stars comes about because the clocks appear to run more slowly in the strong gravitation at the surface of the dense dwarf stars. For surface gravities that are increasingly stronger, the surface clocks run increasingly slower, and for a gravity sufficiently strong the clock would stand still —nothing would seem to happen. Such a strong gravity would result if the object were very much more massive than an ordinary star or if it were compressed down to a very much smaller size, or both. An object having the mass of the sun would have to shrink to a radius of only 2 miles before the surface clocks would appear to stop to an outside observer. That radius, which is proportionately smaller if the mass of the object is smaller, is called the "Schwarzschild radius" after the German mathematician Karl Schwarzschild who first described its bizarre properties. There is another feature of the Schwarzschild radius that is equally intriguing—light emitted from within this radius can never get out, so curved is the structure of space-time around such a collapsed object. No light, particles, signals of any kind can get out—nothing! The black hole is a dark source of gravitational pull into which things can fall, but from which nothing can emerge. It is, therefore, "black" in an ultimate sense of the word. Expectations from stellar theory are that such objects must form from time to time, and astronomers now search for them by looking for the signs of their only influence—that of gravitational attraction. There are some indications that they may have been found orbiting other visible stars, and other indications that huge black holes may exist at the centers of galaxies. I am even beginning to suspect the existence of a small one at the very center of our own sun, devouring it slowly but providing an explanation of the missing neutrinos from the sun.

Continual testing of every conceivable gravitational effect is going on at a vigorous pace. It is fascinating to see if Einstein's theory can withstand the constant probing and remain the correct theory of gravity. No amount of success can ever prove a theory right—only more probable—but a single contradiction proves it wrong or incomplete. Many think the theory is completely correct, but many others think it will prove incomplete or just plain wrong. Certainly history stands on the side of the latter group. It is poetic that the theory resists being cast into a quantum mechanical version because Einstein never did accept quantum mechanics as a complete theory. He dismissed its probabilistic interpretation with "God does not throw dice." Well, maybe He does and maybe He does not. And maybe what is straightforward to a higher mentality just looks like dice to human beings. A great mystery still lurks here.

Whatever one thinks about gravity, one must consider the meaning of its weakness. There are only four forces known to all of physics, and of these, gravity is by far the weakest. One measure of this weakness is a comparison of the gravitational force and the electric force, between a proton and an electron. The electric force is about 10^{40} times greater—that's the number one followed by forty zeroes, a number so big that it defies the imagination. It is equal to the ratio of the size of the known universe to the size of an elementary particle. If the gravitational force rather than the electric force bound the two particles in the hydrogen atom, the atom would be so weakly bound that its radius would exceed that of the universe. That is astonishingly weak. But that very distinction lends credence to Einstein's idea that the nature of gravitation is altogether different from the nature of other forces.

Since gravity is so weak it is interesting that it is so important to the universe and to our own lives, where it doesn't seem weak at all. Laborers lifting stones at Stonehenge would have been understandably annoyed at any deprecation of gravity's comparative strength.

So too would have Newton, who found in it a force strong enough to continuously accelerate the entire earth in its orbit of the sun. I've puzzled on this many times, even after understanding the simple physical answer, for it is poetically like Jesus's beatitudes: "Blessed are the meek, for they shall inherit the earth." The bald fact is that other forces are so strong that we cannot experience them. The nuclear force tightly holds fast moving particles within the atomic nucleus at the center of the atom, where the violence is indescribable—but suffice it to say that the collision energies within a single cubic centimeter of pure nuclear matter would generate enough heat to boil the earth's oceans. But fortunately, only infinitesimally small chunks of nuclear matter, the nuclei of atoms, are found in our world, and they are held in isolation from direct experience by strong electric forces at the center of the orbiting electrons.

The electric forces are themselves so strong that positive attracts negative with such force that it is impossible to assemble a huge number of either positive or negative charges. So strong are the electric forces that they provide the impression of solidarity as they hold the particles fast in their allowed orbits and positions. As I recall my childhood satisfaction in hitting my uncle's steel anvil with a steel hammer, I am all the more impressed to know that that hard resilience was only the electrical repulsion of the electrons and that that great steel strength is only the mutual electric attraction of its ions and electrons. This same fierce force causes the objects to be electrically neutral, a state that can be achieved because the electron and proton have exactly equal but opposite charges. On the other hand, it *is* possible to assemble huge numbers of electrically neutral atoms, as nature has done in the stars and planets. Although the equal numbers of positive and negative charges cancel out the electric force between, say, the earth and moon, the gravitational effects of these particles are additive. Thus, gravity becomes the strongest force between assemblies of huge numbers of particles. The sun has 10^{57} of them! The weakest has become the strongest.

Gravity holds us to the surface of the earth—or to take a more Einsteinian point of view, the surface of the earth prevents us from falling, which is the natural inertial state. It keeps the moon falling along its monthly orbit around the earth, and the earth-moon system falling along its annual orbit of the sun, which itself orbits the center of the Galaxy every 100 million years. The last may sound a long time to us, but it is sufficient to have allowed about forty-six revolutions since the solar system was formed. Gravity holds the gas together in stars so that they may shine, and it clusters the stars together into galaxies. It keeps the earth from flying apart as it spins daily about its rotation axis. The structure of all heavenly bodies in the universe is molded by gravity and rotation, and Einstein suspected the same to be true of the universe itself.

To Einstein it would have been satisfying if *all* motion were relative, if there were no absolute reference frame. It puzzled him that acceleration has apparently *absolute* features, whereas velocity is only *relative*. If a force is applied to an object in space it accelerates according to $F=ma$. Accelerates with respect to what? With respect to the local inertial frame defined by an object falling freely in the gravitational space-time molded by the massive objects in the universe. But what if the object were the only object in the universe? With respect to what could it accelerate? To an absolute space? Einstein, and, following him, I, too, shrink from that idea. Suppose someone had asked Lefty Einstein, my imaginary pitcher, what the stadium rotated with respect to. It's an easy question to us picturing the Astrodome on a giant turntable in Houston, but what if it falls along through space, alone in the universe? Not surprisingly, the unprecedented genius, Newton, had considered it all three centuries before. He proved, or at least he and everyone else thought he proved, that acceleration was absolute. He asked us to imagine a bucket of water spinning on its axis. If the bucket rotates but its water does not, the surface of the water is flat; but if the water is allowed to rotate with the bucket before the bucket is suddenly stopped, the water keeps rotating and makes a concave surface. Thus, Newton argued, since the bucket rotates relative to the water in both cases but the water surface differs, it must be the absolute rotation of the water that determines whether its surface is concave or not. This argument overwhelmed contemporary opposition to absolute motion.

Einstein was not so sure, for absolute motion made no sense to him. The flat surface on the water occurs precisely when the water does not rotate with respect to distant stars. The traditional explanation is that the starry heavens also do not rotate, but Einstein wondered if the distant matter did not determine the concavity of the water in some physical way. If the average matter in the universe determines the inertial properties of matter, then Newton's law $F=ma$ for an isolated object in space might apply to its acceleration with respect to the average motion of the distant galaxies in the universe. Their presence could even *cause* the inertia m of an object. This idea is called "Mach's Principle" after the physicist-philosopher Ernst Mach. It is an appealing idea, but although Einstein was much motivated by it when considering the universe, he never was able to join the idea in a natural way to his theory of general relativity. Phrased more generally, there still is no satisfactory answer to the question: "Does the state of the entire universe influence the value of local physical quantities, such as the mass of an electron, or its electric charge?" There have been many suggestions that it does, but still no quantifiable physical connection has been found.

Einstein's equations of general relativity describe how matter moves in the presence of other matter. In 1917 Einstein began the subject of relativistic cosmology by applying his equations to the universe. To do so he had to make reasonable assumptions about it. He imagined that the universe extends without end, although, because of its curvature, it remained finite. This idea need not be as confusing as it sounds; it is like the surface of a sphere, on which a line can continue forever although the total number of dots painted on the surface (galaxies) is not infinite. Einstein could imagine such possibilities because he had already shown that space is curved. He imagined that the density of matter was, on the average, everywhere the same, and that the motions of the galaxies were so small that they could be neglected completely (a *static* universe). His equations provided an embarrassment, however. When evaluated they demanded that the universe must accelerate! Either it must expand or contract, but in either case, the energy density of matter causes it to continuously change its motion. Einstein could find no solution to the problem. After much trial and error, almost in despair, he offered a modified version of the theory of relativity. It lacked the complete, simple beauty of the original theory, and clearly Einstein did not like it, even though it allowed his original theory to still dominate solar-system phenomena. He introduced a new term with no justification and multiplied it by such a small number λ that it had no effect except on the universe as a whole. In 1917 he wrote:

> In order to arrive at this consistent view, we admittedly had to introduce an extension of the field equations of gravitation that is not justified by our actual knowledge of gravitation. It is to be emphasized, however, that a positive curvature of space is given by our results even if the supplementary [λ term] is not introduced. That term is necessary only for the purpose of making possible a quasi-static distribution of matter, as required by the fact of the small velocities of the stars.

Later he was to admit that it was the greatest mistake of his life, for by sticking to his guns as he had heretofore done he could have predicted that the universe must either contract or expand. For even then astronomers were learning that the stars were clustered into galaxies that move apart from each other. After "Hubble's Law" was announced in 1929, Einstein and astronomers were quick to drop the λ term, which was by then called "the cosmological constant," as unnecessary. Einstein's evaluation of "his mistake" is obviously too harsh because he was only trying to build a theory in accord with what he thought to be the facts about the universe. To not have done so would have been more mathematician than physicist. Enthusiastic astronomers and cosmologists have now leaped too far the other way, hailing Einstein's "prediction" of the expansion of the universe as the most dramatic evidence of the correctness of the general theory of relativity. Analogous results exist in Newtonian cosmology, however, if one assumes an infinite universe of uniform density. Einstein himself reviewed this peculiarity in the introductory remarks of his 1917 paper.

Einstein's theory does provide the most intellectually satisfying physical framework for interpreting the expansion and the associated redshifts of the light of the distant galaxies. Many mathematicians found new solutions to Einstein's equations, which for the homogeneous universe could be reduced to a single second order differential equation for the factor scaling the distance between two points in space. Today one of the major aspects of observational cosmology is the attempt to match the observations of the universe to one of the models satisfying these equations. It is a noble effort, and some astronomers invest a lifetime in the effort. It is much too early to tell if the plan will succeed, but one thing is virtually certain: Whatever humanity's future view of the universe will evolve to, the relativistic concepts that Einstein created will remain an influential part of that view.

Astronomy is built on the techniques for measuring events in the universe. By 1916 Einstein had begun thinking of perhaps the most exotic astronomy of them all—gravitational waves. The basic question is easily enough seen by asking what happens on the moon if the earth is suddenly moved. In the Newtonian era it was commonly assumed that the moon immediately experienced a different force. If that were true, however, instantaneous gravitational effects could be used to synchronize all clocks and the special theory of relativity would then be wrong. Einstein's theory is a relativistic field theory, and the field consists of the curvatures in space-time. When a massive object causing space-time curvature is suddenly altered, the modified curvature propagates outward with the speed of light. These can generate traveling disturbances that may, for example, cause two objects to alter slightly their distance from each other when the wave reaches them. For more than a decade, physicist Joseph Weber has searched for these waves arriving from violent events in the universe. He does this by looking for strains on a 1½-ton aluminum cylinder as it tries to stretch or compress by almost hopelessly small distances when the wave moves through it. It is both puzzling and exciting that Weber sees such events and that they have not yet been "explained away" as due to some other cause. Other workers, however, have not been able to reproduce Weber's results, which contain even a tantalizing indication that these waves come from the center of our galaxy. Einstein would have been surprised, but then he did not suspect that the inanimate universe was such a violent place! Because of the extremely weak nature of gravitation, a detectably strong wave could only originate in a sudden asymmetric motion of a huge mass. Such things now seem barely possible—for example, when a massive star collapses in some nonsymmetric way to nuclear density or even to a "black hole," or when the nuclear crust of a neutron star is shaken by a violent "starquake." The astronomical signs of violence in the nuclei of other galaxies suggest that even larger masses than that of a star may collaborate in producing the necessary cosmic ripples in space. So maybe, just maybe, we may one day have gravitational wave observatories recording evidence of the location and frequency of the most bizarre events known to man. Weber has proposed that an instrumented moon would make the best observatory.

Fate arranged it that I was married in that same city where Einstein's most beautiful thoughts were hatched. We had not chosen Bern for that reason, but we do find the coincidence to be inexplicably satisfying. My wife Annette is German. Driving back on our wedding day from Bern to Annette's Black Forest home, we passed through the town of Aarau, just 14 miles south of the Rhine River, as it flows from east to west along the German border. Here Einstein attended the Swiss Cantonal School, following his self-imposed exile from Germany and the Munich Gymnasium. According to Banesh Hoffmann's wonderful biography, *Albert Einstein: Creator and Rebel*, it was here that the young Einstein sowed the seeds of his theory of relativity by asking what a light wave would look like to someone moving along with it. Annette confirmed readily that she could easily understand Einstein's leaving the German gymnasium because her own high school years in the Schwarzwald were blighted by an aging and tyrannical mathematics professor who shouted "Rise!" when he entered the door and who grabbed those students by the ear who had not studied their lessons. It sounded like a parody to me, but she assured me that it was still real in the 1960s. Einstein was apparently never a good student, in the sense we so often use the word, for he was too motivated by his own curiosity to comply with the recipes of educators. Annette and I were interested to discover in Munich that after the famed Maximilianstrasse crosses the bridge over the River Isar, its name has been changed to Einstein Street by an admiring citizenry. After he relinquished his German citizenship, Einstein obtained his diploma within a more relaxed environment in Aarau in 1896. He was then able to become a student at the famed Zurich Polytechnic.

In reading the harrowing biographical details, I breathe a sigh of relief, as if Einstein's making it to the Polytechnic were a narrow escape for science. His rebellious character could have kept him from formal education altogether, and one wonders if his independent nature would have then been sufficient for his destiny. After those fantastic seven years in the Bern Patent Office, Einstein at age thirty returned to Zurich as a professor at the university. This was the first of a series of distinguished professorships. Fate continually keeps Annette and me in Einstein's shadow, for we fly in and out of Zurich every time we return to her home, just across the Rhine. The whole thing is, to me, wrapped mystically in love, which provides my poetic license for the personal associations revealed here.

One thing seems certain. Science and what it represents was the most precious thing in Einstein's life. He himself said so. I have seen this same passionate love of beauty in almost all great scientists, as in the artist or the musician. Whatever it is due to, it is personally very important. In his biography of Einstein, Hoffmann found relevant these words that Einstein spoke at an official celebration of Max Planck's sixtieth birthday:

> I believe with Schopenhauer that one of the strongest motives that leads men to art and science is escape from everyday life with its painful crudity and hopeless dreariness, from the fetters of one's own ever-shifting desires. A finely tempered nature longs to escape from personal life into the world of objective perception and thought; this desire may be compared with the townsman's irresistible longing to escape from his noisy, cramped surroundings into the high mountains, where the eye ranges freely through the still, pure air and fondly traces out the restful contours apparently built for eternity.
>
> With this negative motive there goe a positive one. Man tries to make for himself in the fashion that suits him best a simplified and intelligible picture of the world: he then tries to some extent to substitute this cosmos of his for the world of experience, and thus to overcome it. This is what the painter, the poet, the speculative philosopher, and the natural scientist do, each in his own fashion. Each makes this cosmos and its construction the pivot of his emotional life, in order to find in this way the peace and security that he cannot find within the all-too-narrow realm of swirling personal experience. . . .
>
> The supreme task of the physicist is to arrive at those universal elementary laws from which the cosmos can be built up by pure deduction. There is no logical path to these laws; only intuition, resting on sympathetic understanding, can lead to them. . . . The longing to behold [cosmic] harmony is the source of the inexhaustible patience and perseverance with which Planck has devoted himself . . . to the most general problems of our science. . . . The state of mind that enables a man to do work of this kind is akin to that of the religious worshiper or the lover; the daily effort comes from no deliberate intention or program, but straight from the heart.

CHAPTER VIII
WHERE ARE THEY ALL GOING?

It was a quick walk from the Kellogg Radiation Laboratory on the Caltech campus in Pasadena to the Robinson Laboratory where the Astronomy Department offices were. On the second floor I walked into Jesse Greenstein's office and introduced myself to his secretary. Professor Greenstein bears the somewhat stuffy title, Executive Officer of Astronomy, and, even more importantly, was an influential policy maker for, and extensive user of, the 200-inch reflecting telescope on Mt. Palomar. In addition, he had more file cabinets than anyone else at Caltech, or so I had heard.

I had just recently finished my Ph.D. thesis and was a postdoctoral fellow in physics at Caltech. I already had had the pleasure of meeting Greenstein on several occasions, largely due to his long friendship with my advisor, Professor W. A. Fowler. It was not uncommon to have the chance of a social evening in the Greenstein's house on San Pasqual Street, about a quarter mile east of Fowler's on the same street. The conversation on such occasions was always good, for Greenstein is a gentle man with varied cultural interests, and his forthright honesty demanded the best from others. But it took no special privilege to see Professor Greenstein; one only had to come with something interesting to say or to ask, especially about astronomy. His secretary waved me on in.

Greenstein's office was large and appealing, as only an office can be that was built in an older and more gracious time. I had heard that Greenstein had enough 200-inch telescope photographs in his file cabinets for a lifetime of research. Knowing little at that time of observational astronomy, I found that thought mysterious. So friendly and direct was his greeting that it was easy to get on with the conversation, even for a relative beginner like myself with a great man of science.

I had just recently had the excitement of discovering a nuclear clock capable of measuring the age of the atomic nuclei of the chemical elements. I came to ask Greenstein about the average degree of ionization of the element rhenium in the interstellar gas because I suspected that the rate of its weak radioactivity could be altered by ionization. We looked up ionization energies for rhenium, and from them Greenstein estimated how much ionization rhenium would suffer in the gas. It was a routine problem for an astrophysicist, and in retrospect I see that I should have directed it first to someone less busy than he. Even a graduate student in astrophysics would have sufficed. But he didn't mind. Greenstein then asked me about some of the new developments in nuclear astrophysics. For about 15 minutes I described the problems I was working on. Before I got up to leave he said, as if just having made a decision: "Wait. I want to show you something." He opened one of his famous file cabinets and pulled out what I recognized dimly as a stellar line spectrum. "I've been puzzling over this for quite a while," he continued, "and I can't make head nor tail of it. See all these spectral lines?" I nodded, without adding that I didn't know a thing about spectroscopy. I realize now that he knew I didn't know a thing about it, and that he probably hoped my ignorance would free me to come up with an original new idea. "There are hundreds of spectral lines here," he went on, "and their wavelengths don't correspond to a single element we know on earth. There is not a single line here that looks like the lines we see from ordinary stars." It crossed my mind that different elements exist elsewhere, or that electrons have different mass or charge elsewhere, or even that the laws of physics are different elsewhere, but I didn't have the courage to suggest such "nutty" ideas. I just shook my head like a fool —thinking as I did that he wanted help with some conventional astrophysical problem of line intensities. I was pretty naive.

Two years earlier something looking very much like a star had been discovered at the exact position in the sky of radio source number 48 in the Third Cambridge Catalogue of radio sources. It was called 3C 48. The first spectrogram of its atomic emission lines had been obtained by astronomer Allan Sandage in 1960, and Greenstein had taken up the challenge of interpreting its undecipherable spectrum. I never asked him later if it were 3C 48 he had shown me, and I don't believe he told me its name at the time; but I feel certain it must have been. It could also have been 3C 196 or 3C 286, which Sandage and Matthews had already located, and it might even conceivably have been 3C 273, whose accurate radio position had just been determined from a lunar occultation in Australia. It doesn't really matter which quasi-stellar object it was, for Greenstein's hope to get a good original idea from me proved futile. I was too excited by my own recent discovery to dream that opportunity could be knocking again. Within the year the puzzle was deciphered. Astronomer Maarten Schmidt, a colleague of Greenstein's at Caltech, discovered that four broad emission lines in the spectrum of 3C 273 had wavelengths in the same ratio as the four common emission lines of atomic hydrogen, but each wavelength was 16 percent longer than in terrestrial measurements. Immediately the spectrum became clear as being one in which every line was redshifted 16 percent. Greenstein then recognized that the spectrum of 3C 48 was redshifted by 37 percent. Such large redshifts were unprecedented in objects that looked like stars. The quasi-stellar objects (QSOs) and their mystery had been discovered. Like the rest of the world, I read about it in the newspapers.

The drama of that fantastic discovery caused me to relearn how the universal redshift had first given evidence of the expansion of the universe. Fortunately for me, it had begun right where I was, in Pasadena, California, during that "decade to remember." Edwin Hubble had been continuously improving the astronomical techniques for measuring the distances to the other galactic systems. His work in the mid-1920s had been an important part of the proof that other galaxies exist. Armed with the world's best telescope, the then new 100-inch reflector on Mt. Wilson, Hubble was quite understandably interested in obtaining distance measures for even more distant and fainter galaxies. He did this by the comparative brightness of objects of known luminosities within the galaxies—especially Cepheid variable stars and nova explosions. They enabled him to estimate galactic distances as great as about 50 million light years—some 2,000 times more distant than the most distant individual stars within our own galaxy. In 1929 he published the astonishing

A cluster of galaxies located far beyond the Hercules star cluster of our own galaxy as photographed by the 200-inch telescope on Mt. Palomar. The galaxies are the various fuzzy objects, whereas stars in our galaxy show up as sharp points—the brightest with crosses that are caused by telescope optics. Spectrum colors of these galaxies are all shifted toward red because of their motion away from us. (Hale Observatories)

conclusion that the more distant a galaxy is, the faster it recedes from us. Commonly called "Hubble's Law," it is written $v=Hr$, where v is the velocity of recession of the galaxy and r is its distance. The proportionality constant H is called "Hubble's constant," in spite of the fact that theoretical arguments suggest it is not really a constant but may instead change in value as the universe ages. This law is regarded by many as the single most important observed fact about the universe as a whole, and it has formed the basis of an extensive plan to try to discover the history of the universe.

Astronomers cannot measure these velocities directly. The only objects one can actually see moving are nearby stars, and then only by photographing their positions at widely different times and comparing. The positions of distant galaxies appear fixed, even on photographs taken decades apart. The recession of the galaxies is inferred instead from the redshift of the spectral lines. The method had been known for a long time for individual stars within the galaxy. Their speeds are typically tens to hundreds of kilometers per second relative to the solar system. Even before Hubble's Law it was known that some distant galaxies appeared to be moving away at speeds greater than 1,000 kilometers per second, but it was Hubble, who showed that these great speeds increased linearly with increasing distance. A speed of 3,000 kilometers per second is, by the way, 1 percent of the speed of light, and the effect on an object moving away at that speed is that the wavelengths of every line in the atomic spectrum is increased by almost exactly that same 1 percent. At such "slow" speeds, the percentage change in wavelength equals the percentage of the speed of light with which the source recedes. Einstein's theory of relativity modifies that for very rapid recession; as the speed approaches closer to that of light, the redshift increases very rapidly.

The discovery rocked the world. "Where are they all going?" was a natural question then, and, indeed, the answer is still not known today. One might also wonder, as Rutherford did about radioactivity, "Why aren't they all gone?" If the expansion continues to increase the separation between galaxies, one day there will be no more to be seen. How lucky then that we are here to see it! It is equally clear that if the expansion has gone on with roughly the same rate in the past, then the galaxies would have been densely packed about 10 billion years ago. Shortly before that time, individual galaxies could not have existed, and one would instead have to imagine a hot dense universe exploding from a state in which future life would have seemed improbable at best. Cosmologists colorfully call it the "Big Bang." Whether a Big Bang actually did occur and, if it did, what it was like is still a subject of hot scientific debate. Still, the incredible thing about the expansion is that it seems to suggest that the universe is an evolving thing, with perhaps both a beginning and an end. Thoughts of this staggered my mind as I relearned this history, and its impact has not lessened in the years since. It is to me a great and wonderful fact more incomprehensible than life itself.

Astronomer Milton Humason began a program to measure the redshifts of the most distant galaxies—galaxies so dim that they cannot be seen with the biggest telescopes but only appear after long exposures of photographic plates. As a result it is now known that Hubble's Law applies to redshifts that correspond to speeds at least ⅓ of the speed of light. At the same time Hubble began a program of counting galaxies to successive limits of distance, and he showed thereby that they are distributed in depth in a homogeneous manner. That is, the numbers of galaxies seem to be, on the average, the same in all directions and at all distances. Like Copernicus, he showed that our galaxy possesses no special location in the universe. It might at first seem special in that all galaxies move away from ours, as if we were at the center of the expansion. That thought is easily seen to be fallacious by considering an infinite raisin cake, which continuously expands in all directions. The distances between all raisins are increasing, and from any one of them it would appear that the others were receding. However, the vantage point of each raisin is identical.

The apparent homogeneity of the universe on a very large scale has led cosmologists to a proposition that is at best an assumption to be tested and at worst a mere hope. The proposition bears the grandiose name, the "Cosmological Principle," and it holds that except for localized irregularities in the distribution of galaxies the universe appears the same when viewed from any location within it. Stated more simply, if some intelligent being on some distant galaxy we cannot even see measures the average number of galaxies in a large volume near him, he counts the same number that we count in the same large volume near us. Furthermore, he could observe Hubble's Law to be true for the galaxies within his view and would even obtain the same value as we for the Hubble Constant of proportionality. In effect it supposes that we see a representative portion of the universe, and other observers see the same thing. The principle greatly restricts the great variety one could have imagined for the universe, and it greatly simplifies any models one might construct of the universe. The assumption also makes Hubble's Law the only possible law of expansion, thereby robbing it slightly of its wonder in return for the security of increased understanding. To see this, one need only contemplate beads attached to a rubber band which is being stretched. The distance between all adjacent beads is equal, and it is maintained so during the expansion by uniform stretching of the rubber band. This distance between all pairs of adjacent beads must then increase at the same rate and—here is the point—the distances between beads separated by one intermediate bead must increase at twice that rate, and between pairs separated by two intermediate beads at three times that rate, and so on. In short, one must obtain Hubble's Law when the beads are viewed from any one of them along the band. There is no other possibility. Thus, in a sense the Cosmological Principle explains Hubble's Law. Of course, it is not known if the Cosmological Principle is a valid and correct assumption, and indeed, we may never know for sure, but substantial evidence now points to its correctness.

Einstein used the Cosmological Principle, without calling it that, in constructing his first models of the universe. Instead of a single source of gravitation, Einstein imagined the universe as uniformly full of gravitating bodies and sought to discover the curvature this would cause on the coordinates of space and time with which we describe events. Unfortunately, Einstein made a mistake (you heard me right) with his own equations of general relativity and constructed what he believed was a proof that the universe had to be static or nonchanging. He then had to modify his field equations by introducing an additional cosmic force. The error in his reasoning was discovered by the Russian mathematician, Alexander Friedman, who showed in 1922 that Einstein's original equations had in fact two solutions, one of which pictured the universe as expanding and the other of which pictured it as contracting. The Belgian theoretical astronomer, Georges Lemaitre, developed the theory further. In 1927 he proposed that the universe started from a highly compressed and extremely hot state, which he called the "primeval atom" but which we now call the Big Bang. As this matter expanded, it cooled, and in a cold low-density state aggregated into stars and galaxies, thereby leaving the complex appearance on the face of the universe. Lemaitre's picture was so influential that he is regarded by many as the "father of the Big Bang"—a rather humorous if nonetheless honorary epitaph.

These theoretical papers were rediscovered following Hubble's proof that the universe actually is expanding. The Friedman classes of expanding models based on the Cosmological Principle and on Einstein's general theory of relativity provided the most plausible framework for interpreting Hubble's Law. In particular, this theory holds that although the Hubble constant H has the same value at every point in the universe at the same time, its value is changing as the universe evolves. A changing Hubble constant means that the rate of expansion of the universe is changing, and one speaks of the "deceleration of the universe" because the theory shows that the expansion rate must be decreasing. It relates the deceleration rate to the average mass density of the universe. Thus, the evolution of the universe in the large is describable by the changing with time of the proportionality constant in Hubble's Law. In fact, the Friedman models are so circumscribed by the theory that measurements of only two quantities today, the Hubble constant and its rate of change with time, would be sufficient to determine the value of H at all times past and future. This possibility presents a great opportunity and challenge to astronomy. To meet that challenge has become, to some, the Holy Grail of astronomy.

The present value of H can rather obviously be determined from the relation $v=Hr$ as measured for distant galaxies. The fact that light travels at a finite speed makes it possible, at least in principle, to also determine the rate at which H changes with time. Light coming from very distant galaxies has taken a long time to get here, and it shows the distant universe not as it is now but as it was when the light was emitted. The most distant galaxy known as a redshift of 46 percent, which would correspond to a Doppler redshift from a source with a speed about 40 percent that of light. With today's optical telescopes, this is about as distant as one expects to be able to see normal galaxies—about 4 billion light years away. That distance means that the light has been traveling 4 billion years in its journey to us! We see the universe there not as it is today, but as it was 4 billion years ago. Thus, by observing the expansion of the distant universe one can hope to measure its expansion rate as it was in the past. In such a way, the rate of change of the Hubble constant could be measured. For example, if the expansion is slowing down, as many believe, the universe will have been expanding faster in the past; as a consequence, distant galaxies will be seen moving away faster than the linear local Hubble Law would have led one to expect.

The trouble with this plan has been that one cannot easily get the needed information. Very distant galaxies are hard to find because they are so dim they eventually become no brighter than the night sky itself. Because they are so dim, measuring the redshift is extremely difficult. To evaluate Hubble's Law one needs the distance to the galaxy. The distance can be judged only by comparing the apparent brightness with the absolute brightness of the galaxy; however, the absolute brightness is itself unknown, and one must rely on "average" galaxies. It is the same problem that had confronted Loÿs de Cheseaux in 1745 when he tried to estimate the distance of the first-magnitude stars. But stars are bright enough to easily study, and a successful theory of stellar evolution relates their absolute brightness to other observable properties. No such theory works on distant galaxies. At the observable distance of a few billion light years, the Hubble constant is not sufficiently different from its present value to easily measure the difference. The net effect of these difficulties is that the study of distant galaxies has not been able to measure the deceleration of the universe. That's a pity because the value of the deceleration will reveal whether the expansion will continue forever to a cold lonely death, or whether it will slow down, stop, and begin accelerating backward in a collapse of the universe. But the quest is not over. With continual improvement of modern electronic techniques rather than the eye and film, and with continued diligence, this noble effort may yet reveal our past and our future. Hubble's words in his last paper are just as relevant today.:

> For I can end as I began. From our home on the earth we look out into the distances and strive to imagine the sort of world into which we are born. Today we have reached far out into space. Our immediate neighborhood we know rather intimately. But with increasing distance our knowledge fades . . . until at the last dim horizon we search among ghostly errors of observations for landmarks that are scarcely more substantial. The search will continue. The urge is older than history. It is not satisfied and it will not be suppressed.

As I quickly relearned this history that had happened just around me I felt a humble respect for the instinctive curiosity of man. I understood the determination of our contemporary colleagues, spearheaded by Allan Sandage, to press on with this quest. And to return to the thread of my introductory thoughts, I could now see why the discovery of the large redshifts of quasi-stellar objects generated such enthusiasm. After 3C 48 and 3C 273, many QSOs were quickly found having redshifts between one and two, which is to say the increase in the observed wavelengths of atomic lines is between one and two times the wavelength observed in the same atoms on the earth. These redshifts were far greater than any that had been previously discovered. It was immediately obvious that if these redshifts were due to the expansion of the universe, the QSOs would have to be easily the most distant objects visible in the universe. Sandage, Greenstein, Schmidt, Burbidge, and the entire world of astronomy—all were thrown into a state of keen excitement. It was like a dream come true—to discover distant objects so bright that they could be seen from almost unbelievably distant parts of the universe. In terms of the Friedman models, QSOs with such redshifts would be seen emitting light when the universe was only about 10 percent of its present age. So it seemed at first that the QSOs would be the key to unlock both the geometry and the evolution of the universe. As usual, however, the path is proving thorny.

The redshift is one measure of distance, at least if the redshift is due to the expansion of the universe. To use it to measure the evolution of the universe, one needs another measure of the distance. For galaxies, one uses their apparent brightness, assuming that the brightest appearing galaxy in one cluster of galaxies has the same absolute brightness as the brightest appearing galaxy in another cluster of galaxies. If that's correct, and it does seem to work pretty well for galaxies, the difference in their apparent brightness is due to a difference of distance. So for the brightest galaxies in clusters of galaxies, the relationship between apparent brightness and redshift conveys reliable information about the geometry of the universe. For QSOs it doesn't work. The apparent brightness of QSOs at a given value of the redshift (therefore presumably also at the same distance) differ one from another by about a factor of ten. Alas, therefore, the apparent brightness of a QSO is not a good indicator of its distance, or at least not a good indicator of its redshift. The idea that the QSO redshift is not due to a great distance has therefore attracted serious scientific supporters.

If the QSOs are in fact not so far away after all, they would not have to be so fantastically powerful just in order to be so easily visible. To be as bright as they are at the distance suggested by their redshifts, the QSOs must put out as much power as 100,000 billion suns! As if that is not incredible enough, that power must originate from a region only about 100 times as big as our solar system—a quite small region by astronomical standards, so small that within our neighborhood it would contain not a single other star. That the QSO is so small is known by the fact that its brightness often changes by a big factor in a few weeks. Thus, its size cannot plausibly be greater than the distance light can travel in a few weeks. The puzzle would in this sense be less severe if the QSOs were closer and therefore not so powerful; but in another sense it would be more puzzling. For what then would be the cause of the redshift? The only two other redshift causes known today are the Doppler effect due to very high speeds of nearby objects and the gravitational redshift due to very strong gravitational fields. But both of these known causes have failed as explanations of the QSO redshifts. If the redshift is not due to great distance, then it is due to some physics still unknown to man. In a search for a guide to the correct answer, a lot of time at the big telescopes is now being spent in trying to find QSOs in physical association with distant galaxies having the same redshift—in which case the redshift is cosmological—or in physical association with nearby galaxies—in which case the redshift is due to something else. To date, possibilities of both types have been reported, so it is still reasonable for a man to take either point of view. But eventually the true situation will become clear. The cosmological cause seems to be winning slightly at present, but, whatever the truth, we can be confident that the question of the QSO distances will be resolved without doubt in the not-too-distant future.

Where are they all going? That question will remain long after the controversy over QSO distances is resolved. Where is the universe going? Astronomers have not been able to find enough matter in the universe to slow down the rapid expansion. Just as a rocket leaving the earth can have too much speed to be stopped by the weakening gravity of the earth, so the universe is expanding too fast to ever be stopped by its own gravitational attraction. It looks like the universe will fly apart forever. If so, galaxies will one day be so far apart that they cannot see each other. Each galaxy will grow darker as its stars burn out. Professor Greenstein is a fan of the poet T. S. Eliot, and he has called attention to these appropriate lines from one of Eliot's *Four Quartets:*

> O dark dark dark. They all go into the dark,
> The vacant interstellar spaces, the vacant into the vacant,
> And dark the Sun and Moon, and the Almanach de Gotha
> And the Stock Exchange Gazette, the Directory of Directors,
> And cold the sense and lost the motive of action.
> And we all go with them, into the silent funeral. . . .

If this vision is of the galaxies, the universe faces death as surely as do you and I. Each galaxy will become a dying ember as its own nuclear fuel is exhausted. Stars will no longer shine. No more distant galaxies will lie within the view of man, who will long since have been extinct anyhow. All is cold and dark and vacant.

I cannot believe it; but I don't know what the resolution of this predicament will be. What I cannot believe is that the universe "only happens once" and that we are just lucky enough to be here to witness it before it is all gone. Perhaps we will find that the universe can slow down, collapse again, and start over; or perhaps we will find that the steady-state, continuously creating universe is true after all.

CHAPTER IX
THE GREAT REACTOR IN THE SKY

The high deserts of the southwestern United States are to me very beautiful—especially in winter and early spring. It was primarily for direct contact with them that Annette and I chose to travel by automobile in December of 1972 to the 141st meeting of the Astronomical Society in Tucson. From the Davis Mountains in west Texas to the Saguarro cactus fields around Tucson the eye encounters the splendor of nature and its most gallant project—life. The desert is full of life in all seasons, life that has worked the most ingenious compromises with its primary source, the sun. The summer sun is harsh, but its great heat also evaporates the oceans, raising fantastic tonnage of water to the circulating currents of the earth's atmosphere, and that lifted mass comes crashing back to earth in precious steady drops of life-giving rain. Just now, the winter sun hangs low in the southern sky, and its rays glance obliquely onto the desert surface, casting long soft shadows of the giant Saguarro even in midafternoon. Sunlight and rain, drought and flood, burning heat and exuberant splashing drops from heaven—the whole cycle depends for power on that great fiery ball.

As our Fiat Sport Spider accelerates uphill, my mind is intrigued by the thought that this motor power also came to us long ago in the soft ultraviolet light from the sun. Absorbed by the chlorophyl, this ultraviolet caused photosynthesis in plants and abundant foliage, and, I have been told, this decaying foliage produced our oil deposits. Even a battery-powered electric car which has gotten its electric charge from a hydroelectric plant ultimately depends upon

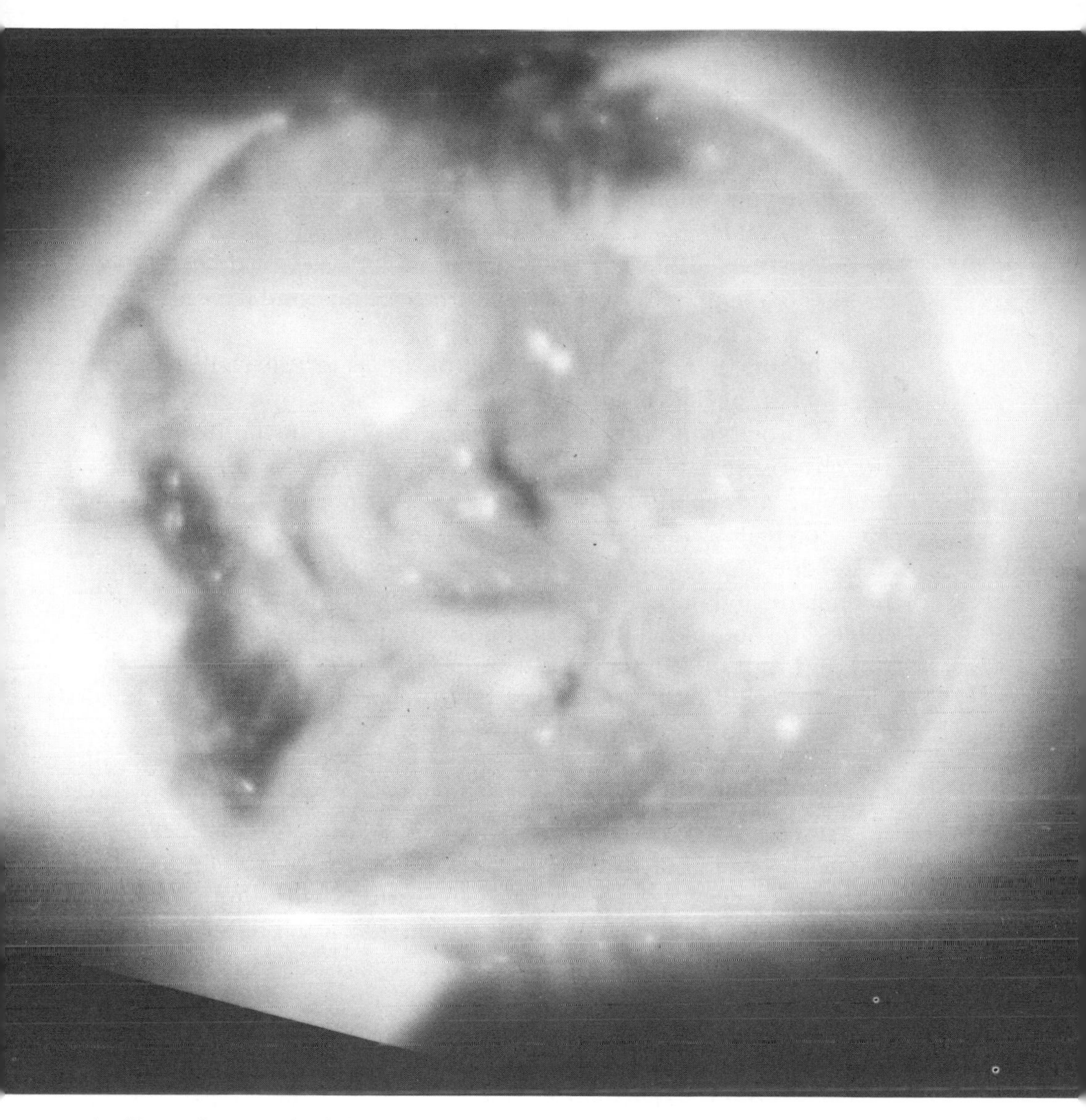

An X-ray photograph of the solar corona from the first sequence of exposures obtained with the X-ray Spectrographic Telescope on Skylab. The observation covers a broad spectral range (3–32, 44–54 A). The active regions, bright spots, interconnecting loops, filament cavities, coronal holes, and other features seen in the photograph are produced by the interaction of the sun's magnetic field and the ionized gas of the corona. The exposure was made at 0336 GMT on 28 May 1973. (G. S. Vaiana)

the river and thereby upon the rainfall generated by solar heat. If the United States receives an average of 20 inches rainfall per year from clouds at 10,000 feet, the same power would put 500,000 million automobiles per year on the top of Mt. Whitney! There's no escaping the fact—without the sun, our planet would be basically devoid of the simple forms of power upon which we traditionally depend. At this time we face an energy crisis of sorts, but I take relief in the assurance that our heavenly provider will not desert us for another 5 billion years. The basic source remains, and it is only up to humanity to rationalize its use.

An historic event of this meeting of the American Astronomical Society is a special session of the solar division, "Observations of the Solar Corona from Skylab." Scientists involved in the planning of the observations of the sun from Skylab are to report on their first findings, and scientist-astronaut Owen Garriott has also come from Houston to describe the observing program in space with the Apollo Telescope Mount. He was one of the crew of Skylab 3 which had been launched on July 28, 1973 for a record 2-month stay in earth orbit. Most of the scientific findings will take time to quantify and develop, but one thing is clear immediately. Skylab confirms the solar surface as a violent place. Dr. Richard Tousey of the Naval Research Laboratory put it picturesquely: "The whole upper atmosphere of the sun is a multiple cauldron boiling away with all kinds of different stews in different places." Astronaut Garriott described huge bubbles of gas within the solar corona that violently burst, throwing out vast amounts of hot solar material into space. These are causes of shock waves within the gas streaming away from the sun, and they have such an impact on the earth's outer magnetic field that the aurorae, or "northern lights," are caused, and radio communications on earth are frequently disrupted. The detailed mechanisms for these things are not completely clear, but the chaotic teeming energy of the solar surface is abundantly evident.

And yet this is but a small portion of the energy put out by the sun. Each square centimeter of earth receives two calories per minute of radiant energy from the sun. The energy comes as 40 billion billion photons of light falling per second on each centimeter of sunlit earth! If a pan of water 1 centimeter deep were to absorb this energy, the temperature of the water would rise 2° centigrade during each minute. Spread over the whole United States, this power is tens of

These three men are the prime crewmen for the Skylab 3 mission. Pictured in the one-G trainer Multiple Docking Adapter (MDA) at the Johnson Space Center are, left to right, Scientist-Astronaut Owen K. Garriott, science pilot; and Astronauts Jack R. Lousma and Alan L. Bean, pilot and commander, respectively. (NASA)

thousands of times the power consumption of mankind within the United States. Viewed this way, the solar power is quite enormous and would be more than enough to satisfy mankind's needs if it could be efficiently harnessed. Within the next decade we may witness giant "solar farms" for collecting the energy from the sun. An area of 1,000 square miles in the deserts of the Southwest could hold enough solar energy collectors and converters to supply the electrical needs of the United States. This may sound large, but in fact it's only a square about 30 miles on a side. There are *ranches* that big in Texas!

But these considerations are still no measure of the sun's fantastic power, for only a small fraction of the sun's radiation even falls upon the earth. Most is emitted in other directions and streams away into space, where it adds its small share to the cosmic reserves. Suppose the earth were only $1/100$th of its actual distance from the sun. The energy striking the earth would then be adequate to bring the earth's oceans to the boiling point in just three days! And even then only $1/100{,}000$th of the solar radiation would be striking the nearby earth. The total radiative power of the sun is 4×10^{23} kilowatts, which is enough energy in a single second to meet the entire energy needs of the United States for 10 million years! The numbers are amusing—even fantastic—and they illuminate a fascinating cosmological question that has haunted me since my career began: What is the source of all this energy that the stars squander into space?

The light photons come from the top layer of the sun's hot surface—a surface that is only about 1 gram thick, for this is as far as a photon of light can travel before being absorbed by the hot solar gases. At the rate at which the sun radiates energy, that surface would cool in a few seconds. It can only remain hot because the sun must be hotter underneath its visible surface. Since heat flows from hotter regions to cooler ones, this subsurface heat could replenish the surface losses. This argument can then be continued, layer after hotter layer, toward the solar center. Enough is known of the physics of heat flow to calculate that at its center the solar temperature must be about 15 million degrees centigrade! That's much hotter than human experience, being about ten times hotter than the highest temperatures ever achieved on earth—namely, for a small fraction of a second in a plasma fusion reactor or at the center of a thermonuclear bomb. One would think that such high temperatures would cause the gaseous sun to fly apart explosively, and indeed it would except for its own enormous weight. The gravitational attraction of this huge ball holds it together, and, in fact, those high temperatures are just what are required at the center to achieve a sufficiently high pressure to support the staggering weight of the overlying layers.

Perhaps it will have occurred to you to wonder what keeps the center of the sun hot. The heat flow outward toward the surface surely cools the interior unless something provides new heat there. About 10 million years would so cool the center that the sun would have to contract due to insufficient pressure in the cooled interior. But that is clearly not happening, for life has existed on earth for thousands of millions of years. The age of the earth is almost 5 billion years, and good evidence suggests that the sun and the solar system formed at the same time. If that be so, the sun has been shining about 500 times longer than it would take for it to cool. The seemingly inescapable conclusion is that some powerful source of energy resides in the center of the sun. Had the Stonehenge men known this, it could only have strengthened their belief in a divine sun.

I first heard an explanation of this in 1957. I was a graduate student attending Professor William A. Fowler's lectures in nuclear physics at Caltech. Not the least of the problems in attending these lectures was that they began at 8:00 A.M. sharp. "Willy," as we astonishingly quickly came to know him, liked to get in a full day's work, and, for him, one of the easiest ways to do so was to have his lectures over by 9:00 A.M. We graduate students, on the other hand, were prone toward late-night discussions. It was a shock to the system to arrive at 8:03, say, and find Willy's large blackboards, which covered three walls, filled with chalked diagrams and equations. Alas, he found it easier to cover large volumes of material by arriving early and writing his notes out ahead of time for all to see. I'm sure we should all have appreciated that, for his notes were displayed with beautiful clarity and compactness, but at the time it only made us wonder how to cope with this professional tornado. Once, as a joke with a message, several of the class spent the night there in sleeping bags and arose, stretching and yawning, when Fowler came in the door. Always the appreciator of a good joke, he carried on with new determination. There, in those circumstances, with voluminous early morning clarity, Willy displayed his ideas that the center of the sun was a giant thermonuclear reactor.

I was to be much influenced over the years by this characteristic of Fowler's that I call *vitality*. His appearance and somewhat robust physique, combined with a jolliness of humor, suggested an easygoing, affable personality. During my years at Caltech he was always ready for a game of softball or football with the students, and on these occasions one could not but notice his agile strength. On mountain climbing trips in later years I was to be impressed with the exceptional physical stamina which he possessed in his late fifties. In fact Fowler loves sports almost as much as he loves physics. He played end on his high school football team and took me to the first Super Bowl in the Los Angeles Coliseum, carefully analyzing how the Green Bay Packers were taking apart the Kansas City Chiefs. He has remained such an avid Pittsburgh Pirate fan throughout his life that he even arranged many of his scientific trips to Washington, D.C., so that they could include baseball stops in Pittsburgh. In later summers in England he would ask his secretary to mail him all the box scores of the Pittsburgh Pirate games from the Los Angeles *Times.* The curious thing about the strength of that attachment is that it is left over from childhood, when his father took him occasionally from Lima, Ohio to Pittsburgh to see the Pirates. Railroad steam engines were built in Lima, and Fowler's continuing love for steam engines still has infectious power. When we were seeking a commemorative gift at the scientific conference in Cambridge, England, celebrating Fowler's sixtieth birthday in 1971, I had the fun of going to London to shop for an antique model steam engine which we presented to him on that occasion. Later, pressed into service as Master of Ceremonies at the banquet for Fowler, I got a really uproarious laugh when, glancing up at the majestic portrait of Henry VIII I remarked: "We are fortunate on this occasion in having a portrait of Professor Fowler hanging on the wall behind him." Those who knew him best laughed hardest! Within a week Fowler located the Cambridge Steam Society, and we were pulled around the miniature track by Willy's puffing engine, running on live steam generated by burning coal. Above all Willy loved a party. Who among us will ever forget the many times he has wagered that he could drink a glass of beer placed on his head without use of his hands, or the times he has divided his face with a mirror, placed a hat on his head, and reduced a young lady to laughter by lifting the hat from his head with the hand hidden behind the mirror when she responds to his order: "Blow!" Fowler's life is a constant stream of such energetic goings on, and on top of this he generates enthusiasm within science and among scientists by his unflagging insight and fervor. He has recently been elected to the next presidency of the American Physical Society, and in the past Lyndon B. Johnson appointed him to the National Science Board.

The ideas concerning the sun as a thermonuclear reactor were certainly not all Fowler's, but he had been involved with them since 1933, when, as a graduate student in Kellogg, he assisted C. C. Lauritsen, a pioneer of nuclear physics in the United States, in some of the early experiments showing that radioactivity was produced when protons moving at about 1/20th the speed of light struck

Margaret Burbidge, Geoffrey Burbidge, William A. Fowler and Fred Hoyle, with Fowler's steam train, presented to him in Cambridge on his sixtieth birthday. (Margaret Burbidge)

carbon. Now Fowler was explaining to us that what actually happened was that the carbon nucleus captured the proton, that is, absorbed it, and became thereby a radioactive isotope of nitrogen. Strange alchemy! And here's the point—that same reaction should be happening at the center of the sun. The sequence of such reactions that occur had been outlined in 1939 in an innovative work by Hans Bethe. Called the "carbon-nitrogen" cycle, it described in detail how carbon and nitrogen nuclei act as nuclear stimulants causing the element hydrogen to be fused to become the element helium. The carbon and nitrogen remain. The relevance of this is that the helium atom, with atomic weight 4, is slightly less massive than four hydrogen atoms, each with atomic weight 1. By Einstein's $E=Mc^2$, the small mass that disappears when four hydrogens become helium reappears as thermal energy in the gas! In this view the solar center remains hot because a small fraction of its mass disappears while hydrogen becomes helium. Only a few thousandths of the solar mass need have disappeared in the entire 5-billion-year history of the shining sun to have provided its power. Even so, so energetic is the sun that 5 million tons of its mass disappear each second!

Lord Rutherford had suggested in the first decade of this century that nuclear energy must power the sun, and Eddington in Cambridge stoutly followed his lead. They could not make a convincing argument, however, for the specific reactions responsible were unknown. Not enough nuclear physics was then known. Bethe's carbon-nitrogen cycle came as a great stimulus because it proposed a detailed path. Earlier with Critchfield he had described another series of reactions in pure hydrogen—that is, without any carbon and nitrogen—that would allow hydrogen to fuse. This series of nuclear reactions, called the "proton-proton chain" because it begins with protons reacting with protons, is now believed to be the major source of energy for the sun. With these ideas, Bethe, a distinguished theoretical nuclear physicist, became the first great interpreter of solar energy. When he was awarded the 1967 Nobel Prize in physics, Bethe was cited especially for this work. His was the first Nobel Prize in astrophysics because as yet there is no Nobel Prize in astronomy.

What I only dimly realized in those early morning lectures in 1957 was that Fowler was the second great man of stellar power. Following defense department work during the Second World War, Fowler and Lauritsen made a conscious decision to devote a major portion of the nuclear research of the Kellogg Radiation Laboratory at Caltech to the study of the nuclear reactions in stars. It is one thing to guess those reactions that can provide stellar power; it is another thing to experimentally measure the rates at which they occur when the particles collide. Fowler, aided by his colleagues and students, is responsible for most of the measurements of these rates.

During my years in Kellogg, every reaction of nuclear hydrogen and helium burning capable of study was studied. The nuclear machinery of the solar center unfolded before my eyes. I remember the many experiments of Kavanagh, Barnes, Parker, Vogl, Seeger, Whaling, Bacher, Tombrello, Larson, Domingo, Davids, and others—all under the watchful eyes of Fowler, Lauritsen, and

Lauritsen. This last famous father and son team has gone since I left Kellogg—both victims of cancer. With this dedicated army and four Van de Graaf accelerators, the conjectures about the centers of stars became numbers, and the numbers became numerical models for the cores of stars. Only the first and most exotic reaction of the proton-proton chains will never be measured. One would have to bombard a hydrogen target with a milliampere of high speed protons for 3,000 years to obtain a single reaction in which a pair of protons would undergo radioactive decay while they scatter one from the other. But in such a way the sun makes heavy hydrogen from common hydrogen, and that is the key to the sun's success. The sun can succeed where we cannot because of the enormous number of collisions within the sun. Each cubic centimeter of the solar center contains 10^{26} protons moving at hundreds of kilometers per second, making some 10^{60} collisions per cubic centimeter per second. Man cannot match this for breathtaking violence, and considering its ferocity it is fortunate that this primary reaction is so improbable—or else the sun would have long ago burned its fuel and died.

The fever infected me. As part of my Ph.D. thesis, I began a nuclear study pertaining to one of the nuclear reactions of Bethe's carbon-nitrogen cycle. Within a year I was staying up late nights to tend one of the small Van de Graaf accelerators that could shed light on these problems. Rubber belts on insulated rotating drums sped continuously carrying an electric charge to a large insulated spherical dome, charging it to potentials of millions of volts. The strong electric field sent nuclei of helium hurtling down through an evacuated tube at speeds approaching $1/20$th that of light. Some millions of millions of helium nuclei each second struck the nitrogen target prepared by Dale Hebbard, a scientist visting from Australia. The roughly 10^{30} glancing collisions each second between helium nuclei and nitrogen nuclei was enough to produce a few thousand nuclear reactions per second, and my counting apparatus was good enough to count about one of these each second. That wasn't much, but it was enough to detect an excited state of the nitrogen nucleus that would, if it had been found to exist, have caused the mass-13 isotope of carbon to be very rare during the carbon-nitrogen cycle. In such a case Fowler and I would have been quite puzzled at the large amounts of carbon-13 that had already been seen on the surfaces of some stars. I must have spent thirty nights, from 8:00 P.M. till dawn, tenderly encouraging the accelerator and apparatus to cooperate and, in the end, was able to show that the threatening state did not exist. It was half a Ph.D. thesis. I was actually somewhat disposed toward more theoretical matters, but Willy felt that, no matter how theoretical my inclinations, an experimental competence was an important touchstone to reality. He was right.

At this point I must acknowledge that being accepted by Willy as a research student was to be the decisive good fortune of my career. Only the admission to Caltech as a graduate student, which made all my subsequent professional life possible, can rank equally in importance to me—especially considering that my bachelor's degree in physics and mathematics came from a modest university in Texas better known for its football teams. Fowler is an easy man to work for. At a psychological low point during my graduate years, his personal concern lifted me up. Willy's knowledge of and passion for his science opened my eyes to the substance of new knowledge, to the paths for making the unknown known. No experience in life is more precious, and those who make its treacherous traverse are stamped for life with its imprint. In this there is no substitute for learning from a man like Willy, who is in the forefront of advancing knowledge. I like to think that I am intellectually independent—a self-made man as it were—but I could not have moved into creative frontiers of astrophysics without the direct, intellectual contact with men who were already there and who showed me the way. As a protégé of Fowler I came to personally know Fred Hoyle, Jesse Greenstein, Tom Lauritsen, A. G. W. Cameron, Geoffrey and Margaret Burbidge, and many others whose names I must omit because my good fortune was too abundant. They are all vivid in my mind—an army of brilliant people in hot pursuit of nature's most fascinating truths.

John Bahcall was in a post-doctoral position in Kellogg at the same time I was—during 1961–1963. He was attracted by Fowler's observation that the proton-proton chains emit neutrinos along with other less exotic forms of energy. Neutrinos are, like light, apparently massless particles that move at the speed of light, but the forces they experience are so weak that they have a very weak interaction with matter. For this reason, neutrinos emitted near the center of the sun come straight out at the speed of light. They do not dally in the sun like the photons do, being continually absorbed and reemitted, continually adjusting their energies as part of the heat slowly flowing outward from the solar center toward its surface where they are at last set free into dark and empty space. Neutrinos are primeval. They emerge directly from the solar interior carrying the same energy they had when they were born a scant two seconds earlier. The sun should emit two neutrinos for every helium nucleus it synthesizes from hydrogen, and they carry away a few percent of the solar power. A very simple calculation shows that 70 billion neutrinos impact each square centimeter of earth each second! They pass through your body and mine in a fantastic bombardment of unfelt radiation. They pass straight on through the earth and emerge on its nightside leaving almost no trace of their existence. Yet they are not imaginary. A few years earlier Reines and Cowan, working at Los Alamos Scientific Laboratory, had detected antineutrinos emerging from a nuclear reactor here on earth. The antineutrino is the antiparticle of the neutrino, and if it exists, so must the neutrino. Although they interact ever so weakly with matter, they do leave telltale traces of radioactivity. Fowler had pointed out that we might find a similar telltale radioactivity caused by the solar neutrinos —especially since a rare branch of the proton-proton chain emits a neutrino of a much higher than average energy.

Bahcall seized on a remark Ben Mottelson made while visiting Kellogg from the Niels Bohr Institute in Copenhagen. These high energy neutrinos could cause chlorine to become radioactive, and with a surprisingly high probability for such a weak force. He calculated almost every conceivable aspect of neutrino production in the sun, and the results were very encouraging. We should be able to "see" the solar neutrinos, and, by doing so, we should see directly into the center of the sun. It should be possible to measure directly the temperature of the center of the sun.

Fowler and Bahcall eagerly pointed out these possibilities to Raymond Davis, who had already expressed interest in building an experiment to detect the solar neutrinos. Perforce, they enthusiastically pointed out these ideas to the funding agencies also because it was no small task to obtain the money for such a difficult experiment. Now the Brookhaven Solar-Neutrino Telescope nests deep in the Homestake Gold Mine in South Dakota. There, 1,500 feet beneath the surface of the earth where it is safe from radioactivity produced by conventional absorbable cosmic rays, a 100,000-gallon tank of chlorine-laden "cleaning fluid" is absorbing the solar neutrinos. Bahcall's first calculations indicated that three to four of those chlorine atoms daily would absorb a neutrino and become an atom of radioactive argon. Davis showed that he could easily find these by collecting the argon atoms every couple of months, but when the experiment began he couldn't find any! He concluded that less than one neutrino daily was being absorbed. A long string of theoretical investigations followed, all of which reduced the expectations somewhat. Current calculations show that the neutrinos should be producing one radioactive argon atom daily. Meanwhile, Davis continues to improve the experiment, and now reports that less than 1/10th neutrino per day—maybe none—are being absorbed! The great reactor in the sky isn't reacting. Our understanding of this god has faltered and our theory of its structure is in crisis, while it benevolently shines away oblivious to our confusion.

But what is wrong? Ours is an age of science and reason, not of myths. These expectations for the sun are all well founded in experimental fact. It is therefore an exaggeration to imagine that astrophysical science will crumble. It seems almost certain that some subtle feature of nature is being overlooked amid an essentially correct scientific perspective. The solar neutrino experiment, like most great and original experiments, is trying to tell us something new. But what? One of the more interesting suggestions is that the neutrino may, during its flight from the sun to us, decay into unknown particles. The flight takes at least eight minutes, even if the neutrino is massless and moves at the velocity of light. Another interesting (and even chilling) observation is that the neutrinos show the rate at which the solar reactor now works, whereas the surface light shows the rate at which heat is flowing out toward the surface. That heat flow requires 10 million years, so it just could be that the sun is now cooling off in one of a series of unknown long-time oscillations. There is geologic evidence that we are slowly entering another Ice Age on earth, and one wonders if the periodic ice ages are related to an unknown pattern of alternating hotness and coolness of the sun. Scientifically the idea is sound, but speculation has not been able to reveal an adequate cause for such slow pulsations of the sun.

I find that this proble of the missing solar neutrinos will not leave my mind, notwithstanding the fact that I have done little to clarify it. I have proposed two possible resolutions of the difficulty, but neither is very convincing. In the first resolution I proposed that the most energetic particles within the sun, which are the ones that produce the neutrinos, are somewhat rarer than is expected on the basis of the simplest theory. My second proposal is that a small black hole (about the mass of the planet earth) rests at the center of the sun. The trouble with these explanations, interesting as they are, is that they have been put forward to explain only one single fact. They really need wider relevance.

The other half of my Ph.D. thesis attacked a subject that has been a constant thread of my research ever since. The question is that of the origin of the atomic nuclei of the chemical elements. It had become a really big part of the Caltech scene in 1953, when Fred Hoyle visited for the first time. In 1946 he had published a paper showing that iron should be the most abundant metal because of the thermodynamic properties of its nucleus; namely, if any elements are allowed to mix at a sufficiently high temperature, such as might be found in old stars without hydrogen or helium, they will turn to iron. Hoyle was increasingly interested in stars much older than the sun—stars that had consumed their nuclear hydrogen fuel and been pressed on to less abundant reserves. It was part of a larger interest in the origin of the elements. Hoyle supposed that the nuclei of the atoms have not always existed, nor that they were created in an instant of time at the beginning of time, but rather that they are in fact synthesized continuously in stars by the same nuclear reactors that provide the power. Indeed the two had to go together physically, for the nuclear energy is provided in any fusion reactor by fusing light nuclei into heavier ones. The next nuclear fuel after hydrogen is gone is the helium that it has synthesized in the process. Hoyle had been thinking of helium fuel. In order that carbon be synthesized at a rate sufficiently rapid that it could be as abundant as oxygen, Hoyle predicted in 1953 that the carbon-12 nucleus possesses a specific excited state with nuclear properties specifically chosen to facilitate its formation in a hot helium gas. Fowler's colleague Ward Whaling guided the experiments confirming the existence of this state, and Fowler later told me that Hoyle's dramatic prediction of it based solely on astronomical evidence was what really "hooked him" on nucleosynthesis. By 1957, the same year I attended Fowler's lectures, he and Hoyle, along with Geoffrey and Margaret Burbidge, published a huge paper for the *Reviews of Modern Physics* presenting a detailed scenario for the synthesis of all of the elements in the evolving nuclear reactors of the stars. In this view the universe originally contained no heavy elements, but has instead manufactured them inside of stars. When the star disrupts at death, it distributes its nuclear ashes into space, where they mix with the gas already there. When our solar system formed almost 5 billion years ago, its gas contained ashes of carbon, nitrogen, oxygen, sodium, magnesium, silicon, sulfur, calcium, iron, and all the rest. In this view, all the atoms of our daily existence (with the exception of the hydrogen in the water) were once long ago synthesized within a now dead star. It was a romantic theory then, and it still is today. And the evidence today is so strong in its favor that it is surely correct.

The major change in emphasis of this theory since those years has been a shift from the slow burning of the apparently constant stars to the violent explosions that occur at their death. Astronomers call these explosions "supernovae." Each of these liberates as much nuclear power in a few ferocious seconds as the sun has liberated in 5 billion years! To the eye one looks as bright as the entire galaxy of 100 billion stars in which it occurs. Most of the new atomic nuclei are probably synthesized in this final flash. That shouldn't be too hard to believe considering our experience with the relatively feeble man-made thermonuclear explosions. The element Californium was, as I mentioned earlier, first found in the fallout from the first hydrogen bomb at Eniwetok in November 1952. It is heavier than any element found on earth and was obviously synthesized in the blast. How much more could be synthesized in these incomparably more violent heavenly blasts! A definite answer is difficult to obtain. One can only calculate, as intelligently as one can.

It is of some interest to sketch the recent years of this project from my personal vantage point. At Professor Fowler's invitation I spent the 1966–1967 academic year on leave from Rice University at Caltech. Back in the environment I knew so well, I had a marvelous time working with Willy and David Bodansky, a professor of physics on sabbatical leave from the University of Washington, on a problem that proved to be of great importance for the origin of the elements. The question was that of how the element silicon, a prominent constituent of sand and glass, would fuse, in a nuclear sense, into heavier atomic nuclei when it was heated. We knew nature had to confront this problem because it was already clear that massive stars build up huge balls of silicon at their centers as they burn their remaining oxygen. With the aid of a key suggestion from Richard Wolf, another Caltech protégé who became a faculty colleague at Rice, we found that the conversion of silicon to iron could be described in a beautiful and powerful way. It showed directly why a dozen or so of the abundant nuclear species between silicon and iron have just the natural abundances that they are observed to have. It established without doubt in my own mind the correctness of the key assumption of the science of nucleosynthesis; namely, the abundances of heavy elements reflect the nuclear properties of those elements and the thermonuclear environments in which they were assembled by nuclear reactions.

You can imagine my enthusiasm when Fred Hoyle, who also came to Caltech in 1966, invited me to begin spending my summers with him in Cambridge. He had just been able to establish there a new institute dedicated to the development of the theoretical foundations of astronomy and cosmology. It was called The Institute of Theoretical Astronomy and had financial support promised over a five-year period from the Scientific Research Council of Great Britain, the University of Cambridge, and various private foundations. It was hoped that the institute, which we called by its acronym IOTA, would also become a planning center for British astronomy. Hoyle wanted to establish a group of American scientists who would come annually, thereby ensuring some continuity in the outside expertise visiting his institute. Eagerly I accepted, along with Willy Fowler, Margaret and Geoffrey Burbidge, and Wallace Sargent. When Willy and I arrived in Cambridge in early May 1967, we toured the construction of the new buildings on Madingley Road and set up our own office, courtesy of Cambridge University Observatory, in a nearby wooden hut amid a field of sheep. It was a pastoral beginning to six consecutive summers of new discoveries in nucleosynthesis—a period that established for Cambridge and for Rice University a special niche in the history of that subject.

The following summer David Arnett came to IOTA for the first time. We were to become close friends and colleagues. During that summer he made a key exploratory calculation that set a major

The Texas Mafia in Cambridge. From the left, Clayton, Schramm, Arnett, Talbot, Howard, Woosley, and Hainebach.

theme of our research during this period. On a computer he heated carbon to a temperature somewhat in excess of that at which it fuses gently, and watched the resulting thermonuclear explosion. I remember clearly his showing me his result—that the thermonuclear debris contained the correct concentrations of ^{12}C, ^{16}O, ^{20}Ne, ^{23}Na, ^{24}Mg, ^{25}Mg, ^{26}Mg, and ^{27}Al (the prominent isotopes of carbon, oxygen, neon, sodium, magnesium, and aluminum) to account for the natural abundances of those species. We are indeed a piece of stardust—exploding stardust! Dave and I began an exciting and productive collaboration on this general problem that Dave called "explosive nucleosynthesis." He came to Rice as a faculty member in 1969, and we published about half a dozen major papers on the research done jointly at Cambridge and at Rice. Ours was, I think, an effective collaboration, in which Dave carried the main thrust of the calculations and the formulations of the astronomical setting, while I snooped around among the calculations for those specific nuclear clues that illuminated the problem. I had excellent Rice research students during those Cambridge years, especially Stan Woosley, Mike Howard, and Kem Hainebach, each of whom accompanied me from Rice to Cambridge on several occasions. Each one of them contributed significantly to the explosion of knowledge that derived from our many calculations of the thermonuclear explosions of different fuels—hydrogen, helium, carbon, oxygen, and silicon. The annual inpouring of Texans into Cambridge even came to be known as the "Texas Mafia." (And Willy was our "Godfather.") Much of the calculational effort was made possible by free computer time on the institute's IBM 360–44 computer. Hoyle had realized from the beginning that the computer is the experimental tool of theoretical astronomy, and he had managed to arrange to have one available on the institute grounds for the free use of institute personnel. Hoyle is one of the great fund raisers in British astronomy.

This exciting epoch in nucleosynthesis has passed, but I was thrilled to have been such a part of it. Hoyle's computer is gone from the institute, as is Hoyle himself, who resigned his professorship and left Cambridge in 1972. I will return to this later. The Texas Mafia has dispersed. Arnett and Howard are at the University of Illinois in Urbana; Woosley is with Fowler at Caltech; Hainebach and Schramm, who had by that time joined Arnett in Austin, have moved to the University of Chicago. We will always remember the scientific adventure we shared during these marvelous years from 1967 to 1972. They are typical of scientific solidification, and the way it so often concentrates into a few key years at some international center of scientific excellence.

On our way back from the Astronomy Society meeting in Tucson, we had to drive on Sunday. President Nixon's request had closed a lot of gasoline stations and in many big crossroads like Ft. Stockton we found not a single station open. Our earthly sources of power were running into trouble. Nonetheless, by tanking in small western towns like Balmorhea and Bakersfield, where we found small independent operators open, we managed to keep going. The naturalness of the winter desert was reasserting itself in my thoughts. It was deer season in Texas, and in the Hill Country half the vehicles were pickup trucks with rifles mounted behind the windows. Texas has a huge deer population, and Annette counted about thirty deer grazing near the roadsides at sunset. They too keep an eye on the sun, and they become adventurous as it sets. Unfortunately we also found a lot of dead deer. Where the highway crosses one draw between Ozona and Sonora, we found eight dead deer lying beside the road. Most were apparently hit by hurrying cars despite the roadside signs warning, "Deer Crossing." We stopped to look at one large buck, and I confirmed that he seemed to have been killed by a collision rather than a bullet. Yet one of his antlers had been neatly clipped away. We were filled with remorse for this once proud creature which was brought down so brutally by man. Its world and ours seemed suddenly so different. It never cared that 70 billion solar neutrinos per second pierced each square centimeter of its skin and, at that moment, neither did I.

CHAPTER

2001: A VIEW OF APOLLO

The history books tell us that the voyages of Columbus and Magellan changed people's view of the world and, indirectly, their view of themselves. That the world is round was already known to many. The Pythagoreans apparently believed the world was round, and in the third century B.C. the Alexandrians actually measured its curvature. Eratosthenes of Cyrene (276–196 B.C.) was the chief librarian at Alexandria and was perhaps the most learned man of his time. He performed one of history's great scientific works by measuring the radius of the earth. He found that at noon on a midsummer's day at Aswan (then called Syene) a vertical rod casts no shadow, whereas at Alexandria 5,000 stadia to the north the vertical rod casts at the same time a shadow of easily measurable length. (The definition of the vertical could be obtained by hanging a weight by a string.) The sun's rays are parallel, so the angle they made with the vertical at Alexandria was also equal to the ratio of the distance from Aswan to Alexandria to the radius of the earth. By such cleverness a small percentage of people came to appreciate the spherical nature of the earth and the associated unity of its surface. But the later and more dramatic discovery of the New World and the ways to navigate the globe are what really changed the life and thought of the average person. It eroded the delusion that the earth extends forever. It thereby changed "natural philosophy" into a meaningful everyday fact. The collective realization that there is no more land to discover—no new worlds to expand into—might have become an important stimulus to effective world government. That it is not yet so can be seen in the limited effectiveness of the United Nations to deal with contemporary international problems and conflicts. In the long run, the necessity of the world living with itself is the important "spinoff" of those voyages of exploration. Of course, for the people involved, the short term payoff of treasure hunting, exploration, trade, empire building (and crumbling), and a large assortment of wars were of more compelling immediate interest.

Are we seriously to believe that the Apollo program will have an analogous impact? Many advocates of the man-in-space program allege that it will, but some of their arguments are nonspecific and not totally convincing. Enthusiastic suggestions that virus studies in space will lead to a cure for cancer, for example, or that the vast and accurate communication network will lead to a better home radio or a TV-phone remind one too much of the circus barker. More convincing is the argument that one can never expect to know in advance what the effect will be of a discovery of fundamental importance on human affairs. Those in favor of this argument usually cite nuclear physics and the voyages of exploration mentioned above. However, it is remarkable that the Manhattan Project and the Magellanic tour only confirmed what the wisest already knew; that the nuclear bomb is possible and that the world is round. Interestingly enough, much less is publicly made of the discovery of unstable "fundamental" particles by nuclear physicists (mesons, hyperons, antimatter) and of the first sightings of galaxies outside the Milky Way by Magellan (the Large Magellanic Cloud and the Small Magellanic Cloud)—probably because these discoveries remain mysteries playing no obviously important role in everyday human affairs. But back to the point, there did exist *individuals* who knew in advance the significance of the Manhattan Project and of Magellan's voyage, although their wisdom had insignificant impact on public thought. This situation is not uncommon. The public has never paid much attention to individuals of outstanding foresight. Indeed, why should they when there are always so many more persons posing as wise individuals and holding the shortsighted view? We simply do not know who the wise ones are. It is *prima facie* ridiculous to expect us collectively to recognize those few individuals whose foresight exceeds our own. Yet even now they live among us, outlining their ideas to the appreciative few. Perhaps the major intellectual problem throughout organized society is that of identifying the genuinely beneficial ideas amid the jangles of conflicting opinion. Usually it is up to history to identify them in retrospect. Perhaps, too, the major virtue of science is not its benefits, but rather that it teaches us so clearly that some ideas are right and some are wrong. Equally important, I think, is the property of science in showing that ideas that seem correct in a limited context of fact may fail badly in the light of new facts. The training of young people in both the problem-solving techniques of science *and* the history of its failures seems to me well advised.

As a practicing cosmologist, I am of course fascinated by the scientific issues that have been illuminated by the Apollo program. My own theoretical perspective has led me to new interpretations of some of the anomalous chemical abundances found in the lunar samples that the astronauts brought to Earth. I have, for example, interpreted anomalous overabundances of two noble gases, argon and xenon, as being due to the infall of cosmic dust that was formed long ago in astronomical explosions. Interpretation of lunar history will certainly depend on whether I am right about that or not. And yet, my thoughts at this moment are not really on the strictly scientific issues to which I am personally devoting my life. It is admittedly thrilling to study the vast quantities of new information relevant to our understanding of how the Solar System formed so very long ago and of events that have shaped the solar system since that time. I also share the belief that the facts about these universally interesting questions will ultimately play a significant part in the human race's cultural consciousness. Yet I also share with the non-specialist a vague uneasiness—an uncertainty of what difference it really makes to life here on earth. I am hard pressed to imagine practical consequences of a technological type, although I do see a beneficial effect on technological industry and overall national competence, and I am aware of certain military possibilities that will probably never be of tactical importance. To those buying food at spiraling prices amid depressing noise and pollution in the cities, such a specialized cultural perspective might be thought to be at best a noble curiosity and at worst a waste. I share that uncertainty insofar as technological consequences are concerned. Magellan's voyage probably stimulated the ship building industry and accelerated the development of astrometrical instruments, but that would have been of little overall consequence were it not for the fact that those events were a part of a world-wide cultural development that was to forever alter the conditions of international relations. I think the Apollo program too must be viewed in terms of a sweeping and crucial change in the way people think.

How shall we discover such a change? We should look for a cultural change rather than only a scientific enlightenment of the few. It must be visible at the level of common activity. It must influence the psyche of large numbers of people. It must be popularized. In short, it must be as visible as a money-making motion picture. There exists such a film, and its impact on me within this context was great.

2001: A Space Odyssey is the motion picture I refer to. Across the dismal canvas of contemporary commercial film, *2001* speaks with a voice as if from a new renaissance—*Thus Spake Zarathustra.* The film's impact comes from its reflection of a psycho-cultural situation of contemporary mankind. It has for the most part defied critical analysis by harried film critics lost in contemporary cliches. It has instead been appreciated by a teenage woman who feels unidentified anxiety; by a bright and idealistic young man who in repeated viewings has come to regard it as part of his need for a deeper meaning for life; and by a middle-aged man who refuses to yield his vitality to the common voyeuristic sexual doubts of his contemporaries. I cannot presume to clarify the artistic meaning of the film. But within the context of these thoughts I think I detect how the film reflects a cultural reorientation induced by the Apollo program.

Many critics have identified the brilliance of the film and its technical excellence. Indeed it is a marvelous thing to see the loving detail of the technical simulation, but it is also quite clear that the impact is much more than that of a technical *tour de force.* Upon reflection I think that the effect of both the film and the Apollo program has been to take the space environment out of the domain of science fiction and to place it into reality—the cautious maintenance of the space-station air, the complicated life-support systems, the artificial gravity induced by rotation, the need to repair any damage to the habitat, the relativity of motion, and the central need for a computer to regulate and coordinate the entire system. These needs have all been suggested to people before, but never have they seemed so starkly natural. Who can forget three Soviet astronauts who returned to earth dead because their capsule leaked. I will not forget swimming with Ed White before their disastrous capsule fire.

Terror lurks. The pulse quickens as in some black and primitive night. There is, after all, a quiet and desperate determination necessary if life is to survive. The film suggests a great hope—that by carefully maintaining the balance of life and by following our curiosity, the age of Zarathustra's "superman," of humankind reborn, can come upon us—*if only we can survive.* By contrast the melodic and seemingly incongruous strains of "The Blue Danube" suggest that the space station is an historical and cultural event. Millions have flocked to see it in their local cinema.

Fred Hoyle, famed British cosmologist, first brought the key idea home to me on January 6, 1970, in the old Rice Hotel in Houston, when he delivered his invited dinner address to the attendees of the historic Apollo 11 Lunar Science Conference. The scientific results of man's first landing on the moon were being debated by scientists at this conference. But in his speech Hoyle reminded the participants of a prediction he made in 1948: "Once a photograph of the earth, taken from outside, is available—once the sheer isolation of the earth becomes plain, a new idea as powerful as any in history will be let loose." Subsequently, in Houston, Hoyle said:

> Well, we now have such a photograph, and I've been wondering how this old prediction stands up. Has any new idea in fact been let loose? It certainly has. You will have noticed how quite suddenly everybody has become seriously concerned to protect the natural environment. Where has this idea come from? You could say from biologists, conservationists and ecologists. But they have been saying the same things as they're saying now for many years. Previously they never got on base. Something new has happened to create a world-wide awareness of our planet as a unique and precious place. It seems to me more than a coincidence that this awareness should have happened at exactly the moment man took his first step into space.

I marvel at the correctness of this observation. Hoyle was, as far as I know, the first person to identify this "spinoff" from the Apollo program. Many years ago when I first read Rachel Carson's *Silent Spring* I was moved and angered, but I shared the then common attitude that she was a romantic. The *wise individuals* of the 1960s spoke unheeded of the escalating dangers of pollution, in spite of the fact that both the pollution and its damaging consequences were quite evident. Times have now changed. In Stockholm, the United Nations in 1972 conducted a historic international conference to

Earthrise behind the moon (NASA)

attempt to begin cooperative international programs to resolve many environmental dangers. That was a small step for each delegate but a big step for mankind! Our green and blooming spacecraft is, after all, a spacecraft. We now listen seriously to the reasons for various predictions concerning the fate of life forms of this planet, but we suffer from the inevitable inability to recognize today's wise individuals. Because we don't know who is right, there is a real chance that we cannot act in time.

But, one thing was very clear to me. The money invested in space research was among the best and noblest investments by the United States and Soviet Union. Public doubts about the desirability of all types of space exploration could reflect, I am afraid, the doubts of a great and decisive people who have lost their way. On the very evening on which I began writing this chapter, December 6, 1972, I and millions around the world watched the launch of *Apollo 17* on television. While sympathetically hoping for the safety of Astronauts Cernan, Evans, and Schmitt, it was hard for me to realize that this was to be the end. After their return the great adventure that will have been Apollo was to be over. It will live on in the telling of it, however. Unless we find a good way to continue this quest, folk songs and poets will, before another decade passes, speak wistfully of the time when human beings walked the moon and viewed their earthly home. Skylab and the space shuttle may be the way to continue, but perhaps much less expensive automated spacecraft will do the job better. The important thing, I think, is that mankind in general must share awareness of its new findings, just as we shared the Apollo footprints on the lunar soil. What difference does it make to everyday life? Awareness of the spacecraft, Earth, may be the difference between life and death. By nourishing the curiosity of human beings to know the truth and their primitive urge to expand, we are unwittingly taking part in perhaps the only adventure sufficiently grand to jar us from our suicidal complacency.

In his new book, *Carrying the Fire*, Michael Collins, the *Apollo 11* astronaut, leaves us with this haunting thought: "If I could use only one word to describe the earth as seen from the moon, I would ignore both its size and color and search for a more elementary quality, that of fragility. The earth appears 'fragile,' above all else. I don't know why, but it does."

CHAPTER XI

A GARDEN FOR LIFE

The cosmological eye sees geologic formations not as commonplace features permanently embossed on an unchanging earth but as temporarily frozen clues to the violent upheavals of evolution. Today the South American continent continues its slow drift away from the African continent from which it split off, but on the western shores of South America the Andes are rising in magnificence as the Pacific Plate beneath the ocean floor crashes into and under the crumbling continental plate. As exciting as the earth is, it must now share the stage with the sun's other planets as new facts about them are discovered. And as if to repay the favor, this knowledge from afar gives the commonplace features of our earth a new fascination.

Along the southern border of Texas and the northern border of Mexico lies the Rio Grande, one commonplace feature that I know very well. It has flowed there imperturbably for millions of years, carving out its almost 1,000-mile route to the Gulf of Mexico. Trees and brush and abundant foliage blossom near the banks through hundreds of miles of arid highland desert. In summer the town of Presidio is often the hottest place in the United States, but it survives comfortably with the help of the river. Yet, the river is a geologic instability. Gravity draws the runoff from the spring rains to the lowest points of the terrain, where the mounting flood mercilessly pursues even lower ground. Mindlessly and tenaciously it seeks the sea from whence it came. The rushing floods wash out the lowlands, making them ever deeper and smoother at the lowest points. After hundreds of thousands of years the river route is well formed, and after millions of years it is a thing of grandeur. It is a sculptor of stone and soil.

Heat flows constantly out of the earth from an interior that is kept hot by several causes; it was perhaps initially hot and has not yet cooled, heat is still being slowly released by radioactive decays within, and gravitational settling and crystallization release heat. The massive ball that is earth spins once per day and must preserve its angular momentum even as it settles and crystallizes. Sometimes the pressures and strains are too great. The earth trembles as its crystalline surface cracks, and huge areas are slowly lifted up by great pressures underneath while others slowly sink, and here and there volcanoes burst through and release their pentup fury. The heat then flows out as molten stone. Just west of Presidio, such an uplift began maybe 30 million years ago on land in which the river was entrenched. Slowly the land rose, granites and sandstones and dirt, but it rose too slowly to dam up the incessant river. The spring floods kept coming, and to not be stopped in its course the river washed an ever deeper trench through the rising land. The stone walls rose 1,000 feet above the river. Finally the uplift stopped and the river burst out into flat land to continue its old path. One walks today at this point into the Santa Elena Canyon. The breathtaking beauty of its walls startles the mind—"Time and the river flowing . . ."

From there, the river turns southward to avoid a 3,000-foot uplift that was capped by a huge volcano. Today this old crater basin, or caldera, and its eroded rims are one of my favorite spots on this earth. Its walls still rise almost 8,000 feet above sea level. This extinct volcano, called the Chisos Mountains, is the center of the Big Bend National Park. The river takes the big bend around the Chisos Mountains only to cut its way through another splendid canyon over 20 miles long at Boquillas. The tortuous routes continue eastward through canyons past Langtry, where Judge Roy Bean declared himself the only law west of the Pecos. About 20 miles further east the Pecos River cuts through its own great canyons to join the Rio Grande. Finally freed at Del Rio, the river flows lazily through the greening Rio Grande Valley, where great citrus orchards grow. It is a symphony of life on Earth—the desert, the sun, the rain, the river, the valley, and the sea.

One of mankind's greater curiosities concerns whether this drama

The east entrance of the Santa Elena Canyon of the Rio Grande. The left canyon wall of the river is in Mexico; the right, in Texas.

of life is played elsewhere in the universe. That question found strong expressions in religions and in literature, and it gives psychic meaning to man's first steps into space. At the same time the manned Apollo program was getting under way, the National Aeronautics and Space Administration began an adventuresome series of unmanned spacecraft. Bearing such romantic names as *Explorer, Surveyor, Mariner,* and *Pioneer,* they set out to gain information about interplanetary space and about the planets themselves. The first questions concerned the physical environments of the planets. The most important early discovery was that interplanetary space is not empty. A hot ionized gas flows outward from the sun and passes the planets with supersonic velocity. It came to be called "the Solar Wind," and it affects significantly the outer atmosphere of each planet, but in widely differing ways. Still, the most dramatic findings have been at the planets themselves.

Mariner 9 reached Mars in November 1971 while I was on sabbatical leave in Cambridge, which, although a center of cosmological thought, was nonetheless far from the direct contact with the space-probe teams. Like any curious person in England, I kept abreast of *Mariner's* fortunes in the London *Times* and in *Nature,* the incomparable English magazine devoted to scientific discovery. The novels by Edgar Rice Burroughs about Mars and the romanticized astronomical studies of the astronomer Percival Lowell, who claimed to see an intricate grid of canals on Mars, would at last, if all worked well, be replaced by a scientific knowledge of Mars. *Mariner* 9 was placed in orbit around Mars on November 13, 1971, from where it functioned for almost 12 months and sent back almost 7,200 television pictures of Martian environs. Disappointingly, the first photographs revealed only a seemingly featureless planet, and I feared that either Mars was uninteresting from that altitude or the television cameras had lost their resolution. Fortunately the experts realized almost at once that Mars was engulfed in an enormous dust storm as violent winds were filling its atmosphere with Martian dirt. Yet even then, through that murky atmosphere emerged the peaks of several enormous mountains. That was exciting because mountains are a sign of past geologic activity. One of these mountains rises about 15 miles above the surrounding floor, about three times the height of Mt. Everest above the oceans of the earth! One area, called Nix Olympica, had already been noticed by astronomers with tele-

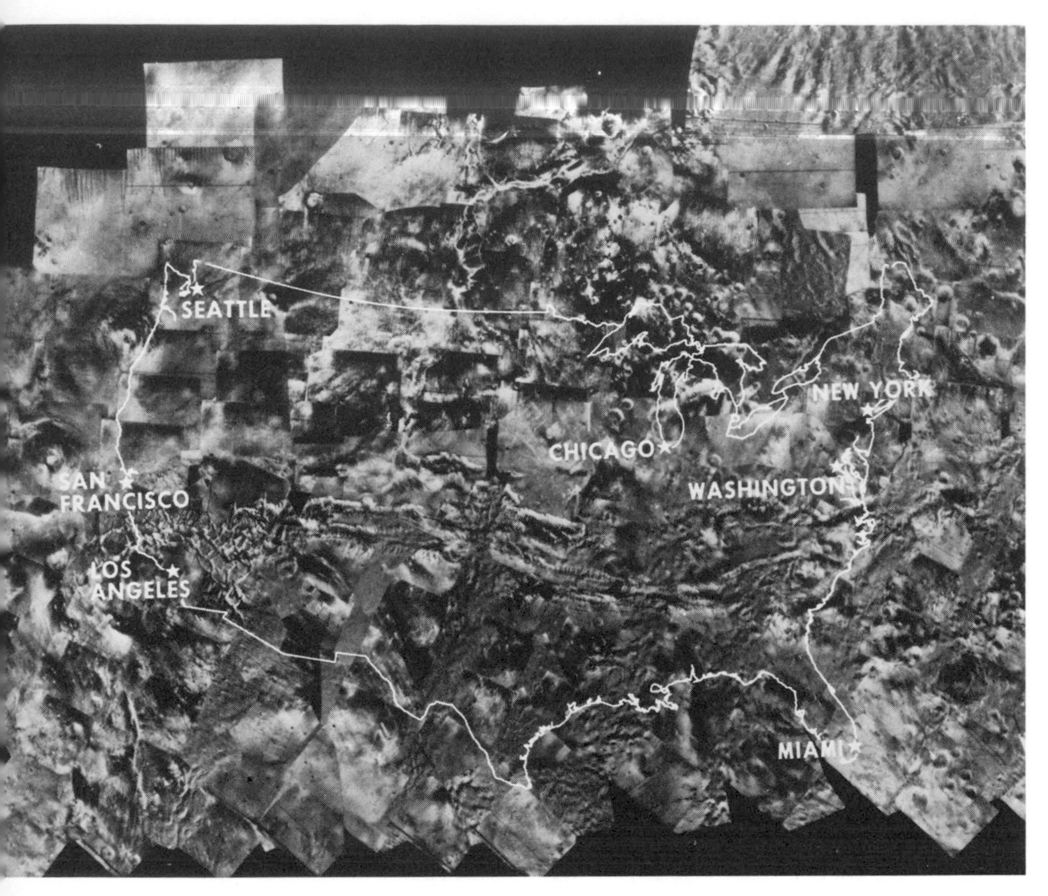

A panoramic view of the equatorial region of Mars is seen in this mosaic of pictures taken by Mariner 9 from late January until mid-March, 1972. Several hundred individual TV photo frames were scaled to size for this preliminary composite picture, which extends from 30 degrees North Latitude to 30 degrees South Latitude and from 10 degrees West Longitude at the right edge to about 140 degrees West Longitude at left. The photo map stretches more than one-third the way around Mars and covers an area of about 11 million square miles, or about one-fifth of the planet's surface. The equator bisects the mosaic horizontally. Upper left portion shows the complex of giant volcanic mountains and their summit craters. The largest of these—Nix Olympica—which measures 310 miles across at the base is in the upper left corner. At least nine huge volcanoes have been pinpointed in the Mariner 9 pictures. The center section of the mosaic contains an enormous canyon—2500 miles long, 75 miles wide and nearly 20,000 feet deep. On Earth, this feature would extend from Los Angeles to New York, with Los Angeles and San Diego on opposite rims. Its depth is three to four times that of the Grand Canyon in the western United States. (NASA)

scopes here on earth, though its exact nature was unclear. As the dust cleared *Mariner 9* sent back vivid pictures of a huge volcano. The Chisos Basin at the Big Bend is nothing compared with the central crater of Nix Olympica. The caldera is 315 miles from rim to rim! It is larger than the huge volcano which forms the Hawaiian Islands. For a cosmologist the planet is definitely not cold and dead. Perhaps, as on earth, those volcanoes belch forth gases that in time become atmosphere and ocean. Perhaps, like here, one may find a garden for life.

Mariner 9 found that clouds containing water crystals appear to form around these large volcanoes. Maybe the water comes from the volcanoes, as it did here on earth; however, it seems more likely that the water vapor comes from ice at the polar regions. The first sign of spring on Mars is the retreat of this north polar cap. The hope for life excites me as I contemplate these Earthlike signs.

The atmosphere must have oxygen, because the ultraviolet spectrometer on *Mariner 9* found ozone in the Martian atmosphere. However, the amount is small. Early in the summer there is no detectable ozone in the atmosphere. In late summer and fall it begins to appear over the polar cap in association with the so-called polar hood. In winter the maximum amount of ozone is found from latitudes of 45° north to the pole. As summer approaches, it disappears again. This sounds full of promise for life—but that same keen *Mariner* tells also the bleaker side of the story. The total atmosphere of Mars is no more than 1 percent of the atmosphere here on Earth. This is much worse than at the top of Mt. Everest. The atmosphere is so rare that it dashes all hope that water in the free state can exist on Mars. It would evaporate. The water on Mars is frozen in the ground by temperatures that reach −190°F near the poles. The prospect is a rather pessimistic one.

And yet, there is a network of tributaries apparently flowing into the Grand Canyon of Mars. The canyon is ten times longer and three times deeper than our magnificent one in Arizona. Were these tributaries formed by water that flowed from time to time? Is Mars simply caught in the grips of an ice age? If the planet were warmed, would melting and evaporating water cause a greenhouse of warmth that would bring torrential rains and floods? Who knows. Incredibly enough, one cannot yet understand what has caused the ice ages here on Earth. It seems certain that the study of the planets will

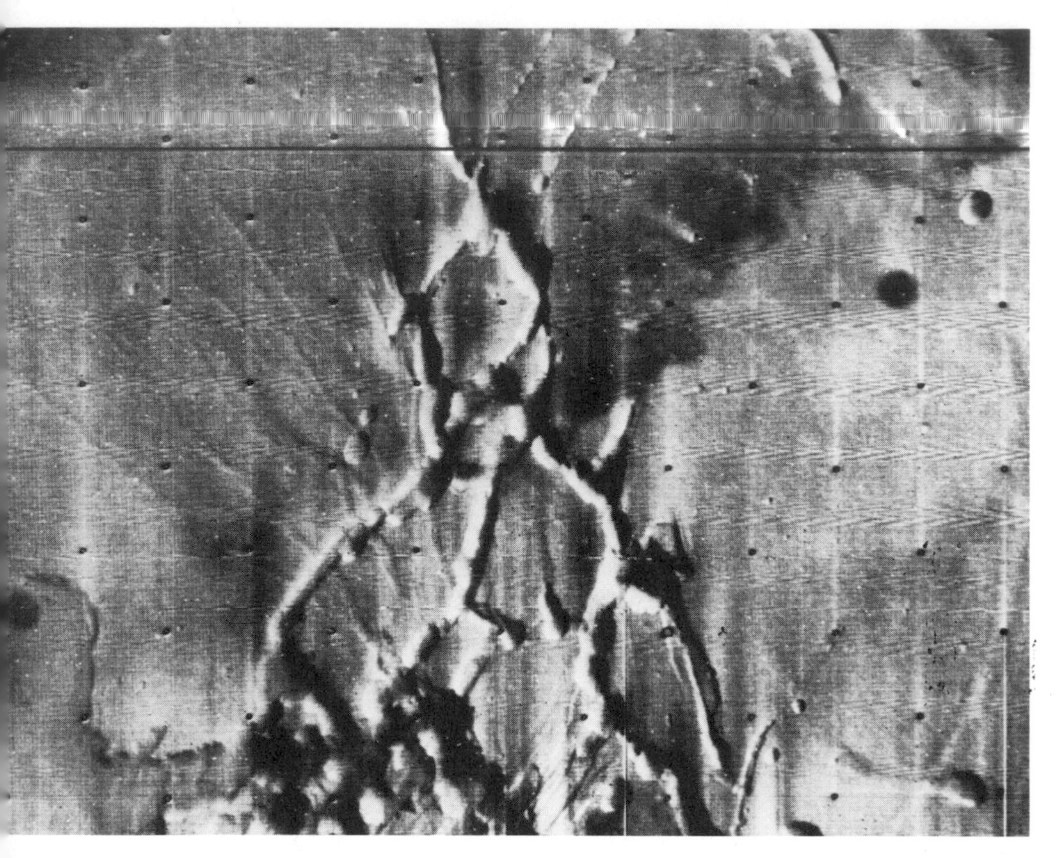

An intricate network of mighty canyons appears to hang like a giant chandelier from the Martian equator in a picture taken by Mariner 9 on January 10, 1972. This photo, which covers an area 542 kilometers wide by 426 kilometers high (336 by 264 miles), provides dramatic evidence of erosional processes at work on the fractured volcanic table lands of Mars' Noctis Lacus. (NASA)

bring great rewards to our understanding of weather because they provide vastly varied conditions that are not attainable here on Earth.

The *Mariner 9* photos show that half the planet is made up of pockmarked plains. Impact craters are very numerous and greatly resemble those on the moon. The impact basin Hellas is larger than any found on the moon. That these portions of the surface retain their craters probably indicates that they are the oldest Martian terrain. Mars shows great variety, resembling Earth in its variety much more than it resembles the moon. And that's fitting. For by

At right is an 80 by 40 mile dune field of loose material in the floor of a 93 mile wide crater in the Hellespontus region of Mars, photographed by the Mariner 9 spacecraft's narrow angle camera. The dune field appears as a black spot, at left, in a wide angle photograph of the crater (arrow). The field is composed of numerous long dunes spaced about a mile apart. The lee slopes of the dunes appear brighter in this picture. As in dune fields on Earth, dunes at the margin of the field are smaller than in the central parts. The similarity of size and direction of the individual dunes indicated they were formed by strong winds blowing from a consistent direction, which in this case is from the southwest. Other craters on Mars also contain dark spots, some of which may be dune fields. Evidence of wind erosion is seen in many Mariner 9 photographs, but this picture is one of the few examples of very large scale depositing by the wind. Mariner 9 has been in Martian orbit since November 13, 1971, has returned 6824 photographs, and has mapped 85% of Mars. The straight dark lines across the pictures are missing lines of picture data which can be restored by a computer process. (NASA)

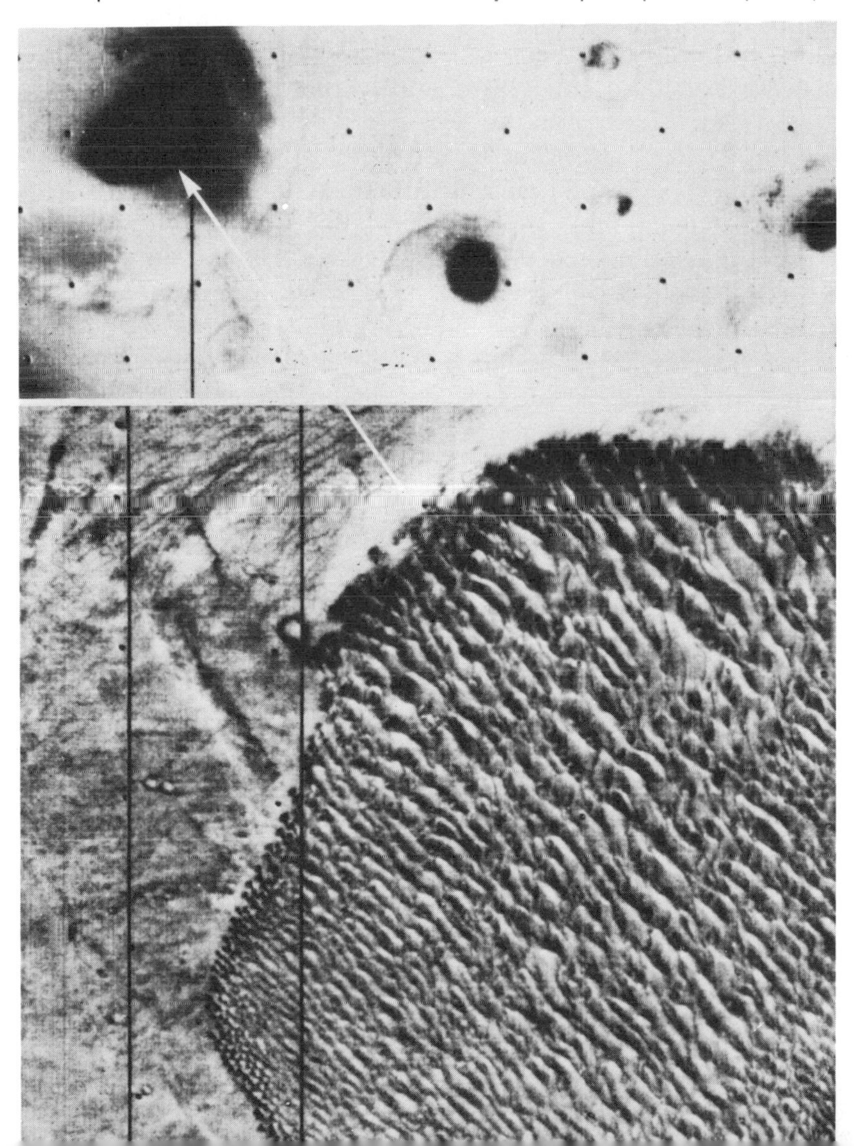

one of nature's freak facts, Mars has the same length day as Earth, and its rotation axis is tilted by the same 23° as Earth's. Stonehenge on Mars would be much like Stonehenge on Earth—but Mars has two rocky and potato-shaped moons, Demos and Phobos, that orbit through the Martian heavens. Could anyone, or anything, have watched their motions in that dark night sky? I find it hard to imagine that any beast could be hardy enough for life to advance to the point of reasoning intelligence there. But then, I *am* limited to my awareness of life on Earth. So I am breathless in anticipation of 1976 and 1977, when two unmanned Viking spacecraft are scheduled to be landed on Mars—that is, unless we panic about the small cost of that adventure.

If Mars seems bleak and cold, Venus is, by contrast, a hell in heaven. While I was a graduate student at Caltech I first heard that Venus had just been discovered to be a bright source of radio waves with wavelengths of about three centimeters. The waves looked like heat radiation, and, if so, the surface of Venus seemed to be hotter than the hottest oven—at least 800°F. This was considered an astonishing result, and I can remember listening wide-eyed as many fine scientists openly expressed the belief that the interpretation must be incorrect. But it was not. By 1968 some of the dreams of science fiction began to come true. The American spacecraft *Mariner 5* flew by Venus and indicated a very dense and hot atmosphere. The Soviet spacecraft *Venera 4* attempted to land on the planet, and it radioed its last message before it ever landed when the pressure reached twenty times that on earth and when the temperature reached 450°F. Later landings of Russian spacecraft, the first successful landings on another planet, show the ground temperature to be about 900°F. The atmosphere is so thick that its weight produces the enormous ground pressure of about 1,400 pounds per square inch. That's like being 3,000 feet under the earth's oceans —deep enough to easily crush any submarine. The pressure had crushed *Venera 4* before it ever reached the ground. This enormous pressure results from Venus having almost a hundred times as many atmospheric molecules as Earth's atmosphere. This dense atmosphere provides a blanket of insulation, like a greenhouse, that allows the surface to become unbearably hot. Light cannot penetrate through this blanket without countless scatterings, and even the sun would be invisible in the dull reddish glow at its surface. Day and night would be virtually indistinguishable, and they might as well be, considering that, for some unknown reason, the planet rotates so slowly that its day is about eight months long. If that description is not frightening enough, consider the fact that Venus's clouds appear to contain concentrated sulphuric acid.

How can it happen that these two planets nearest the earth have atmospheric conditions in such opposite extremes to our own? It is both sobering and thrilling that life in advanced form as we know it here could probably not survive on either planet. I am tempted to think how lucky we are, until I realize somewhat sheepishly that we wouldn't be here to feel lucky unless we had had the right garden to grow in. For human beings and the countless life forms that support them, the earth has indeed been the Garden of Eden. But why are our neighboring planets so different? Did Venus once have a small cool atmosphere until some change caused it to become a hot greenhouse, vaporizing all volatile things into that crushing polluted atmosphere? Could that happen to us? I think we should regard that as a serious question and be cautious with our atmosphere. If supersonic transports are delayed while we estimate their effects on the ozone concentration of the stratosphere, for example, that's no great loss. But one thing is sure—a sudden and substantial heating of the earth's atmosphere could conceivably lead to a Venusian disaster. If more water evaporates, the earth might get hotter yet, vaporizing the oceans, followed by the seeping of carbon dioxide from the carbonate rocks of the crust of our earth. Suddenly we would have a heavy thick stifling CO_2 atmosphere and inescapable heat. My natural instinct is to smile with the thought that "it couldn't happen to me." But carefree glibness chokes on one bald fact—it happened to Venus, and no one yet knows why.

I do know that very mysterious things have happened in this Garden of Eden of ours. I will never forget my first visit to the Smithsonian Museum of Natural History in Washington, D.C. There I saw the reconstructed skeletons of those fantastic dinosaurs that abounded in North America during that great Mesozoic era. There before my eyes were genuine bones, many displayed in stone just as they had been discovered. Natural models set in display windows show the surroundings as naturalists have envisioned it. Most of these large plant-eaters lived in or near shallow water, with abundant natural foliage to feed their enormous carcasses and walnut-sized brains. Their world was a giant swamp, and the thing that so fascinates me was that this giant swamp included Kansas

Colorado, and Wyoming, which are rich in dinosaur fossils. Those swamps are obviously gone, and so therefore are the dinosaurs themselves. After 100 million years of abundance, their world changed forever some 60 million years ago. It seems probable that the central states were lifted up by internal pressures, draining the swamps into the sea. Those poor dumb beasts never stood a chance. In elementary school I saw pictures of the great Mammoth, frozen in ice with vegetation still in his stomach. Even the North Pole seems to have once been more livable. These great misunderstood changes in the history of our planet suggest new possibilities for its future, and I personally would like to understand what has happened.

One approach toward understanding that seems to have great popular appeal is that of the religious or romantic visionary. Moses (or the real author of *Genesis*) thought he saw the beginnings of Heaven and Earth with God's help, and countless millions afterward have zealously defended his vision. The author Velikovsky imagines from ancient texts that Venus was pulled out of Jupiter by strange cosmic forces about 1500 B.C. Velikovsky, a Russian-born physician, is now a rich man from the proceeds of his successful book *Worlds in Collision*, an interesting piece of science fiction that is advertised as nonfiction. His visions have gripped the public's imagination. In Canada, in the summer of 1974, an international symposium entitled "Velikovsky and the Recent History of the Solar System" was held with the express purpose of debating ideas related to the Velikovskian hypotheses. Few physical scientists attended because we have an annoyed impatience with discussions of ideas that fly in the face of the facts. This annoyance has never made scientists very popular with the public, which often prefers to believe that its own fantasies about great and mysterious things should be at least as interesting as the facts. *Newsweek* (February 25, 1974) observes that such ideas "have a special appeal now, at a time when interest in simple explanations of the complex, particularly ones tinged with mysticism and the occult, is at a new high." People have waited in long lines to see the film *The Exorcist*, whose central theme capitalizes on this very phenomenon.

I wish to confess that I am more fascinated by the facts. I care more that the photos that *Mariner 10* is just now sending to Earth, as it flies by Venus, show evidence of raging winds screaming around Venus at speeds as great as 200 miles per hour. The seemingly strong March winds are now gusting across the Texas plains, but Earth after all rotates 243 times faster than does Venus. Why is Venus windy? I cannot climb that volcanic caldera in the Big Bend without thinking of Nix Olympica on Mars and of the ancient millions of years of planetary stress that burst them forth. I cannot look at the exquisite blue of the west Texas sky without the simultaneous joys of understanding its cause in the quantum laws of scattering of light by atoms and of imagining its contrast to the dull reddish opaque glow of the Venusian surface. I cannot wade into the Rio Grande within the Santa Elena Canyon without a puzzlement over a history that gave us abundant free liquid water, and I cannot see those sharp stone walls chiselled by its flow without remembering the *Mariner 9* photographs of the gorges emptying into the great Coprates Rift Valley of Mars. I cannot see the flaming red of a Texas sunset spread across the western sky without thinking that on Mars, if the dust is not blowing, the same sunset shows a brilliant yellow ball in a black sky and that, on Venus, sunset is meaningless and undetectable amid the omnipresent red. As I see the great constellation of Orion twinkling above me on a winter night, I cannot but realize that there, too, are suns—and probably planets. My conscious life has acquired from such facts what is for me the richness and beauty of a fine diamond. The commonplace has become the spectacular. Every aspect of nature holds connections to the rest of the universe in my own mind. It is what Carl Sagan has called the "cosmic connection" in his very fine book of the same title. This sense of unity and knowledge has transformed my life from something mundane to something exciting.

From the 5,000-foot-high Chisos Basin on a black November night, Annette and I watched Jupiter with astounding clarity. Incredible as it was to me, Annette said she saw moons around Jupiter with her bare eyes. I had thought it impossible, but with my 7x35 binoculars, I saw three moons with great clarity in the orientation Annette had described to me. My bare eyes could not see the moons at all, although by rapid blinking I frequently perceived a disklike brightness having the same orientation as the plane of the orbits of the moons. Jupiter has, in fact, a rich harvest of thirteen moons, although only four are easily visible. Those four were discovered on January 7, 1610, when Galileo first turned his newly invented telescope toward Jupiter. The largest Jovian moons, named Io,

Europa, Ganymede, and Callisto are all almost Earthlike in size, having diameters of thousands of miles. The same is true of Titan, the largest moon of Saturn. All five have atmospheres. Titan's atmosphere is much greater than that of Mars. They also have ice. They may have life of some sort. These large moons can exist because the central planets about which they orbit are so large. Jupiter is 317 times more massive than the earth. When it was forming, Jupiter was well on its way to becoming a star when the solar nebula ran out of free gas as all the remainder was trapped in the sun. Even so, giant Jupiter radiates four times as much energy as it receives from the sun, so it has some considerable internal sources of energy. As Annette and I watched it that November night, I could not forget those facts, and I was filled with wonder at what a strange place it must be. So varied are the circumstances there that it seems to me quite likely that some form of life may exist there. I hope I live to know.

That exploration too has begun. On December 4, 1973, after 21 months in flight, the U.S. space probe *Pioneer 10* passed by Jupiter within 80,000 miles of its surface. On its way *Pioneer 10* had determined that the asteroid belt did not necessarily present a hazard to craft passing through it. Rapidly rotating Jupiter has a strong magnetic field, and upon penetrating it *Pioneer 10* found intense belts of trapped radiation like the Van Allen belts above Earth. The equipment suffered severe radiation damage in passing through it, but it is healing itself and is now expected to work until 1977 when *Pioneer 10* will have passed the planet *Uranus*. New information on the moons of Jupiter was also obtained. The density of Io is 3.5 grams per cubic centimeter, much like that of Mars and much greater than that of Jupiter itself. New questions are being asked for *Pioneer 11*, now following *Pioneer 10* toward Jupiter. A *Mariner* mission to Jupiter and Saturn is hoped for in 1977. Not only are these grand and quickening to the pulse, they are also inexpensive. The cost of all of them is much less than was the cost of one week of the seemingly interminable war in Vietnam. Sagan has stated it dramatically: "A decade-long program of systematic investigation of the entire Solar System would cost as much as the accounting mistakes on a single defense weapons system in a single year."

There is one other aspect of *Pioneer 10* that stays with me. It carries a message from mankind. Probably no one will ever get the

Pioneer 10 was 1.6 million miles from Jupiter when it took the spectacular picture above, showing the Great Red Spot and the shadow of the moon Io. (NASA)

message, but at least it is there. It is etched on an aluminum plate attached to the antenna support struts of *Pioneer 10,* where it should remain for hundreds of millions of years of drifting through space. Even now *Pioneer 10* is being accelerated like the crack of a whip as it rounds Jupiter, so that it will eventually leave our Solar System. From there it would move along slowly through the vastness of interstellar space. It is not likely that it will ever reach another Solar System, but perhaps someone will one day detect it in space and fetch it. The message shows an image of a man and a woman, an image of the *Pioneer 10* spacecraft, and a mathematically coded description of the location in the Galaxy from whence it came. It is not a serious attempt at communicating with extraterrestrial life. It is only an afterthought to a planetary probe that will leave us forever. But it's the thought that counts.

There are almost a million million stars within our galaxy, the Milky Way. How many of them, I wonder, have intelligent creatures looking outward from planets in orbit about them. Are there any that look like human beings? Did astronomy also play such a big role in the development of their intellects? Do they teach their children about God, wage wars over territory, and wonder if other intelligent life exists? It seems sure that they are not aware of us, in any case. From even the nearest star to us, the sun would appear as an inconspicuous first-magnitude star and its mighty planet Jupiter would be 100 times fainter than the faintest object we can see on Earth from the Palomar 200-inch telescope. By the same token, we cannot see planets around other stars even if they exist. Nor would I expect anyone to have detected our many radio emissions—either radio stations or dispatched signals to space probes. Our technology in this regard is so recent that the signals, even if they were strong enough to detect, will have traveled only to nearby stars a few light years away. Our civilization is a new-born babe that has only for a few decades had the basic capability of transmitting information. Other civilizations would no more listen to us than a scholar would listen to a new-born babe in the neighboring house.

We have ourselves listened for extraterrestrial communications, and I think we ought to do more of it. The first attempt was organized by Frank Drake at the National Radio Observatory in 1960. It was called project OZMA, and it was not pursued very diligently. A radio telescope listened to two stars for about two weeks, with negative results. But it would be possible to hear if someone has spoken. At Arecibo, Puerto Rico, we have a 1,000-foot-diameter radio telescope that would be capable of communicating with another one like itself anywhere within our galaxy. The question is where to look, and what kind of signals to look for. With more thought about the logical procedure, and with dogged persistence, we may one day succeed. It would be worth it! I have in fact often wondered just what it would mean to me to know that

The 1000-foot radar telescope of Arecibo Observatory in Puerto Rico lies in a shaped limestone bowl among the hills. It could communicate with a replica of itself anywhere else in our Milky Way Galaxy. (National Astronomy and Ionospheric Center)

another civilization is sending out signals bearing intelligent communications. I'm not sure of the answer, but I am sure that it's one of the few questions whose answer holds very great importance to my concept of myself.

If we could be sure that those billions of stars possess comparable numbers of planetary systems, the answer would then be almost certain. If life evolved naturally on Earth, as all but those holding an earth-centered religious perspective believe, then it will surely have evolved also on large numbers of those billions of planetary systems. If it has not, we would have to conclude that an extremely special mystery surrounds us. I doubt that that will be the case, but I would like to know.

The possibility remains that the Solar System may itself be very special. To know the answer to that we must first learn how it formed and how those billions of other stars formed. Information about the many objects in the Solar System is now coming in at a breathtaking rate. Deciphering these clues will take longer, but we may confidently expect a new level of understanding of the origin of the Solar System to emerge from these facts. Even that, however, may not give assurance of other planetary systems. The trouble is that most stars form in double or triple systems, and the process of planetary formation may be fundamentally different when two stars orbit each other than it is in the case of the single sun. Studies of nearby stars show that more than half are visible doubles or triples. And of those that appear single at first glance, most may have a very dim starlike companion, like a Jupiter but many times more massive. The evidence of those massive satellites can be seen in the small wobble they cause in the positions of the observed stars as they and their satellites orbit about their centers of mass. Perhaps these most common stars, the double stars, do not easily form stable planetary systems. Many of the remaining singles are either too bright and do not live long enough for life to develop on their planets, or they are too dim and do not provide a large enough zone of warmth and sunlight. Sensible speculations lead, in fact, to many different conclusions, and no one has been able to present an overwhelmingly convincing argument either for or against the common occurrence of planetary systems. Rather than speculate, it seems best to gather more facts and to study them with all the assembled sophistication of physical science.

A film based on Erik von Däniken's book *Chariots of the Gods* opened in Houston amid unbelievable promotional efforts. I will see the film out of curiosity, although I found the book to be a piece of trash. Every conceivable unexplained artifact of mankind is there put forth as evidence that the earth was visited by astronauts from advanced civilizations. None of it stands up to rational thought. It is the archeological version of the UFO hoopla. There is a lot of money in people's curiosity about their cosmological origins. This fact has for centuries fueled an enormous astrology industry, which ranks with palm-reading and tea-gazing in the scale of man's endeavors.

But its prosperity does, after all, reveal an indomitable fascination with matters of ultimate meaning. This I share. I only hope it does not sour the public taste for the true nature of the scientific adventure. I myself continuously find the facts and their explanations to be far more fascinating than the wildest romantic yearnings.

Up the streamhead, we,
Round Slioch's straggled pawns,
Meall Riabach and *Meall Each*,
Toward her rooks and battlements.
Straining in the climb
Nonseeing in the windswept fog
We mistook the penultimate peak
for that long awaited top.
Squinting at the puzzle
With Ordnance Map and compass
We discovered East was not the ridge
And found the way on up.

For what? Cold rain and wind
Beating our faces without care?
Tired shivers surveying reality
Rather than whiskey and warm fire?
Aye! Comprehending all of that
We go again each time
With aching legs and lungs
A tribute to the vital force.

Down the cautious ridge,
A stop at *Sgurr an Tuill Bhain*,
The final exultation till we
Plod back toward the world.

Clayton

CHAPTER XII
WITH FRED ON SLIOCH

Fred Hoyle and I had driven from Cambridge to Blackpool for an overnight visit with his son Geoffrey at the home of Geoffrey's fiancée. Fred enjoyed this not only because Valerie is a pleasant young lady, but also because her father is another mountaineer. Fred's taciturn summation was "He's a fast man up a mountain." That was as good a way as many others to appraise a man. The next day we stopped off in the Lake District to have morning coffee with another climbing crony and followed it with a quick 2,300-foot climb of Blencathra, one of the beautiful gentle tops in the Lake District, just east of Keswick. It was just a warm up for the highlands of Scotland, but it was near where Fred would be buying a new home within two years.

One of the memorable aspects of my annual summer visits to the Institute of Theoretical Astronomy in Cambridge has been the trips to the highlands with Fred Hoyle. This association is one of the privileges of my life. Hoyle has for 25 years been the world's deepest and most influential thinker on cosmological matters. In this century he stands in line after Einstein and Eddington, and the personal insight into his thoughts has been for me exhilarating.

Partly because he was born and raised in rugged Yorkshire, Hoyle's values reveal innate appreciation of the outdoors—from the rocky boyhood cricket pitch, where defending one's wicket was an exacting task, to the moors and mountains, and ultimately to the galaxies. If that sequence seems simplistic, do not discount its truth. Fred jokingly made me an honorary Yorkshireman, considering that

Hoyle studying an Ordnance Survey Map on the slopes of Blencathra

I *am* a Texan, and I understand his analogy well. The dark night sky there left its impact on him. To Fred nature remains real and raw even in its most distant and esoteric domain. A man of many thoughts but few words, he has little patience with artificially contrived human structures—either physical or intellectual; for example, in his popular brief book *Of Men and Galaxies* at one point he notes: "Walk into a big cathedral, and it wipes your mind clean of every thought." That's Fred. At the same time he is influenced by his awareness of the finiteness of the human brain as an instrument of cognition, recognizing with grim resolve our inability to operate meaningfully outside its particular mode of logic. Driving northwest from Inverness he was offering his thoughts on how the theoretical structure of quantum mechanics might be related to the philosophical problem of *free will* and to the cognition of *the passage of time* when, unfortunately, we were slowed to a snail's pace by two autos pulling camping caravans. About three separate times Fred shifted his Lotus-powered Cortina into second gear, only to find the frequent curves too risky for passing more than one caravan at a time. Muttering "The fools," and other assorted evaluations of their oblivion to traffic, he quit hoping for courtesy and began using his horn. To no avail. Several miles passed before the road straightened enough that we easily passed them, unrecognizing and uncaring in their pursuit of "the roadside outdoors." You see them by the dozens, having their tea or coffee by the roadsides.

From there to Achnasheen, where we were booked for our stay, Fred released his annoyance by suggesting ways to keep Scotland from becoming an amusement park. Even in such circumstances his ideas have fresh penetration. I am still convinced by his argument that improved and widened roads do not improve traffic in places like the Scottish highlands; to the contrary, they only increase the volume of traffic, encourage idle motorists, and aggravate the congestion in the bottlenecks like small towns or curved hilly portions of the roads. Besides, the destinations cannot handle the increased flow. Boarding Boeing 747s in New York or London shows the principle to be applicable there too, and ignorance of it was a major cause of error in the development of the city-suburbia schism in the United States. I now find myself almost completely opposed to the construction (or "improvement") of all new roads because I can never witness it without remembering the devastating Hoyle arguments of the bad side effects. It always reminds me of the rigorous *second law of thermodynamics,* which guarantees that one cannot run an engine from heat energy without dumping large amounts of heat energy into the environment. For Fred it was a long harangue, but for me it was an education.

I was reminded that the earliest published Hoyle paper that had an influence on me concerned thermodynamics. In 1946 he had published a paper in which he argued that iron was the most abundant metal because the atomic nuclei of iron were more tightly bound collections of protons and neutrons than were the nuclei of other elements. Both the high iron abundance and the fact that the mass-56 isotope of iron is much more abundant than the mass-54, mass-57 and mass-58 isotopes was argued to have occurred because a thermodynamic equilibrium gives statistical preference to iron. With no dependence on nuclear physics aside from structural information, he had been able to conclude that if one could heat other metals (titanium, chromium, nickel, zinc, etc.) to temperatures of a few billion degrees and wait, these metals would transmute to iron with the isotopic composition we know here on Earth. I have always been impressed with the far-reaching consequences of this simple argument. The associated ideas have never been far from my career, although interestingly enough, I had only recently come to believe that the thermal equilibrium "cooks" nickel instead of iron. In the anticipated stellar explosions the nuclei are not in my own view provided with enough neutrons to become iron. The nickel produced is radioactive and decays to isotopes of iron in just the proper abundance ratios after it has been expelled from the star (which Hoyle imagined as the thermonuclear "oven"). But enough of that.

The next morning we drove the one-lane road from Achnasheen to Kinlochewe, where we parked the car near the hotel. With a small back-pack of lunch, compass, and Ordnance Survey maps, we hiked in along the Kinlochewe River to the edge of Loch Maree, one of the many highland jewels carved by prehistoric epochs of glaciation and sliding ice. Threatening dark clouds hung over Slioch, one of the picturesque highland mountains, as we began the 3,200 foot ascent to its top. We worked along the edge of the two steep rock formations, named *Meall Riabach* and *Meall Each*, until we found the main ridge to the summit. The wind and rain gave a less glowing view than usual, but the sense of surveying the natural universe, of grasping it and recognizing it, always presents the same reward. It is the life blood of the cosmologist. Casting ones eyes over the tops in all directions, one shares the plight of the astronomer who sees the universe at a moment of time and wonders what it was before. There it all stands, unmoved and unchanging, before our eyes. Only the fantastically small fraction of a second that the light ray spends in traveling from the horizon to our eye separates the observed time of the horizon from that of our own summit, while the whole tens of millions of years of these changing highlands lie hidden in its glaciated clues. The astronomer too sees many stars in our galaxy as they were 10,000 years ago when the signals began their journeys to our telescopes, but that is but as a fleeting second in the tens of billions of years that the cosmologist seeks in his rock garden of clues.

In 1948, in Cambridge, Hoyle described a radical interpretation of the history of the universe. He proposed that on the average the universe appears the same today as it always has. The average distance between galaxies, the brightness of the night sky, the average chemical composition of the universe—all remain constant, so that the universe has neither beginning nor end and so that our own moment of existence is not otherwise special in the grand history of things. The main immediate obstacle for the theory was that the universe is known to expand, and two distant galaxies continually increase their separation. So how can the average distance between galaxies remain constant? Only if new galaxies are created to fill the increasing empty spaces. To accomplish this within the framework of existing knowledge Hoyle introduced a new interaction, "a creation field," into Einstein's equations to describe this creation of new matter. The "steady-state theory" as it is called has great philosophic appeal. It preserves a space-time symmetry in the description of the universe, for it is otherwise normally contended that the universe is (on the average) the same at every point in space at the same instant of cosmic time, but that it is not the same at different times. Such an asymetric model is called an "evolving universe." For philosophers the steady-state has the appeal of rendering irrelevant the problem of what has gone before the early stages of the evolving universe. The expanding evolving universe must also confront the unpleasant position that the universe may expand to a cold death with galaxies beyond sight of each other—and yet here we are alive and in sight of other galaxies. The position is uncomfortable, if not treacherous. Hoyle confronted this problem straight on, and, as so often happens to great ideas in science, saw his paper rejected by two prestigious physics journals before it was published by an astronomy journal in 1948. Both the *Proceedings of the Physical Society of London* and the *Physical Review* (United States) declined the paper, and by the time the *Monthly Notices of the Royal Astronomical Society* published it, they also published in the same volume another paper on the subject by Bondi and Gold.

Like any strong assumption, the steady-state assumption leads to many consequences. It predicts more things than any other cosmological model, which is a great credit of the steady-state theory; but as a result, it has been subject to energetic attack from astronomers. That is as it should be because factual contradiction of any prediction of a strong theory can decisively reject the entire theory—at least in its simplest form. Such a rejection amounts, paradoxically enough, to an unequivocal increase in knowledge, whereas a noncontradictory fact only makes a theory more plausible. Of course, scientists occasionally find a less noble pleasure in proving a competitor to be incorrect. Certainly the steady-state theory has been under continuous scrutiny and withstood several premature and erroneous attempts to discredit it. Curiously enough, the strongest opposition came from within Cambridge itself, where Hoyle held the Plumian Chair of Natural Philosophy—but more of that later. On the next day, Fred was describing to me and to Willy Fowler, who had by that time joined us by one of his beloved trains, a central dispute in current observational cosmology. We were on our way to climb Liathach, an especially splendid mountain on the shores of Loch Torridon, while Fred alternately cursed caravan drivers and described an alternate point of view to the current one regarding the radio source counts.

The question of the spatial density of radio sources is a key one in cosmology because it can reveal the geometry of the universe. Ever since Einstein formulated his general theory of relativity, we have come to believe that the density of matter and energy in the universe affects its geometry—and by that we mean its intrinsic geometry, or its physical curvature, rather than some special feature of the particular coordinates and clocks used to describe events. The distinction is easily made by imagining some two dimensional surface, like a plane, a surface of a sphere, or a paraboloid of revolution. The plane is flat even though curved coordinates could be used to locate points on its surface, whereas the sphere is intrinsically curved no matter what coordinates one uses. Imagine creatures living wholly within such a surface, where all events occur on the surface as if the other dimension did not exist. Light rays from galaxy to galaxy would follow the shortest path, which is a straight line on the plane but a curved line on the sphere. Those creatures would not know that the shortest path on the sphere is curved because their whole experience is confined to that surface, but they could tell that their universe had physical curvature by measuring the angles of a triangle. On the flat surface they sum to 180°, but on the spherical surface they sum to more than 180°, and the excess measures how curved the surface is. Small spheres have greater curvature than large spheres, which become almost flat if they are large enough. So it is also with the universe, except that we experience three spatial dimensions so that the curvature is hard to visualize. We must instead resort to some geometrical test, such as summing the angles of a triangle. Unfortunately, that particular test would require an intergalactic triangle, and we are not in communication with colleagues at the other corners of the triangle.

Hoyle had been involved in attempts to determine the intrinsic curvature of our universe before. In 1958 at the Paris Symposium on Radio Astronomy he explained how measurements of the angular diameter of radio galaxies as a function of their distance (or redshift) could reveal this curvature. Fred explained it again to us by asking us to imagine that we live at the North Pole of a two-dimensional spherical universe (imagine a model globe), and that all light rays coming to the pole have to travel in the curved surface along the lines of longitude ("great circles" or "geodesics") passing from the object to the pole. That is, the light rays travel along the shortest curved distance. Then imagine the apparent angular size of an iceberg as it drifts away from the pole. Its angular size is, of course, the angle between light rays arriving from opposite sides of the iceberg—in this case the angle at the North Pole between the two lines of longitude touching the outside edges of the iceberg. An unusual thing soon becomes apparent; when the iceberg is near the pole, the geometry is almost flat and the angular size of the iceberg decreases inversely with the distance of the iceberg from the pole —just as in Euclidean plane geometry—but as the iceberg nears the equator, the angular size decreases much more slowly, decreasing not at all at the equator where the lines of longitude are locally parallel to each other! Near the equator, say, between the Tropics, all icebergs have almost exactly the same apparent size regardless of their actual distance (assuming the icebergs have the same physical size) from the North Pole. A surveyor with tools to measure this effect could tell that his universe was the surface of a sphere and measure this radius if he knew the distance to the iceberg. Hoyle proposed exactly this technique to measure the curvature of the four-dimensional space-time of the universe. It is pointless to try to imagine this curved space-time in the manner one can imagine the two-dimensional sphere. No one, not even Einstein himself, could visualize that, and we must be content with a mathematical description. However the mathematical cnsequences are analogous to those that happen on the surface of a sphere. Hoyle proposed that the physical size of the newly discovered strong radio galaxies might be roughly constant, so that their apparent angular size as a function of their redshift would give the required information. He pointed out that in the popular "closed models" of the evolving universe, the angular size decreases to some minimum value (the iceberg at the equator) and then begins to increase for more distant objects. Very distant objects look bigger as they get farther away! To see this one need only trace the lines from an iceberg near the South Pole. The minimum angular size of a radio galaxy in this type of universe should occur near redshift $\Delta\lambda/\lambda=1$, and we easily see strong radio sources to distances greater than that. In the steady-state theory, however, the angular size continues to decrease for increasingly distant objects, but it decreases more slowly than in Euclidean space. This brilliant test is proving difficult to carry out because the physical sizes of radio sources are not exactly equal, so that their relative angular sizes are somewhat ambiguous. In addition, it is difficult to obtain their redshifts. Nonetheless, the recent results from the National Radio Observatory in West Virginia do not show angular sizes to start increasing again for the largest redshifts, so, unless something tricky is going on (as is likely), this evidence does not reveal the universe as closed and evolving. The geometry looks almost Euclidean! Of course, the distant sources could be of different physical size than the nearby sources. Either the linear dimensions of the radio sources decrease with the redshift in just a way as to cancel the geometric effect (a cruel trick of nature), or the geometric effect is small (as in an Euclidean or a steady-state universe).

The conversations are always interrupted during the climb. Trudging upward over rocks and through peat beds is better for just thinking. Besides, when climbing over 1,000 feet per hour one is well advised to save one's breath. The east end of Liathach is a steep rock face, where we climbed sharply from 1,000 feet to 2,800 feet. The bulk of the mountain is composed of highly sculptured Red Torridonian Sandstone, which gives it a rosy tint. These famed sandstone beds were laid down in shallow seas 750 million years ago.

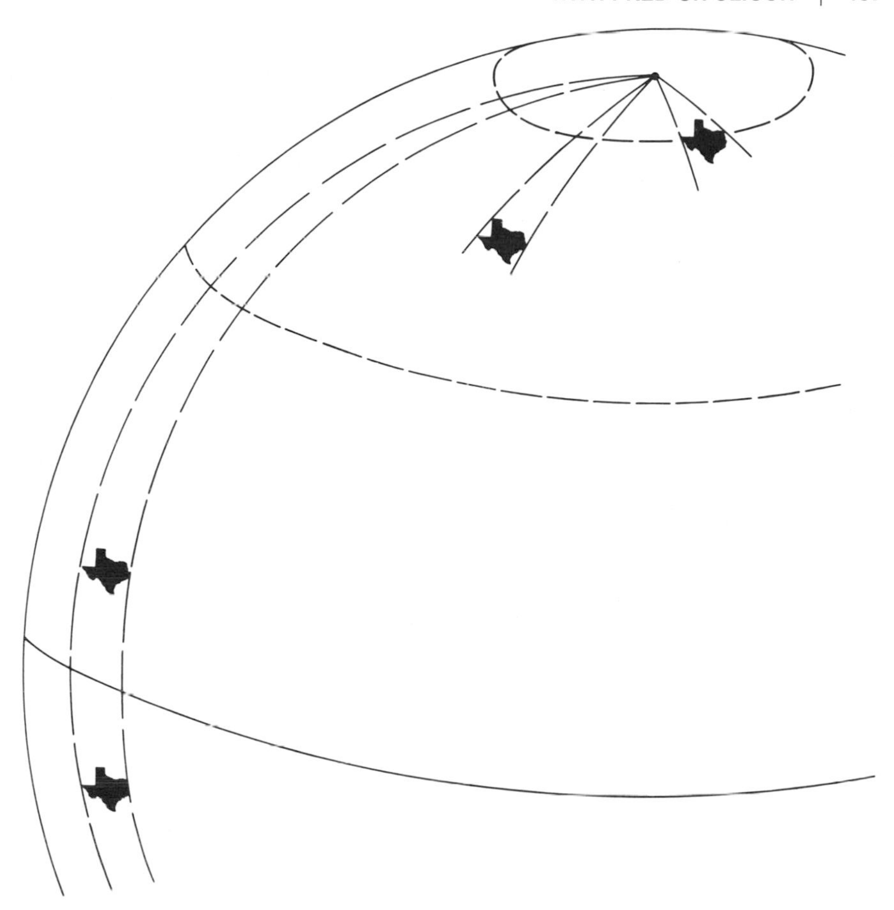

The angular size of an iceberg drifting away from the North Pole decreases with distance, just as in flat space; however, near the equator the angular size no longer changes, just as in spherical space.

Once on top, we followed a ridge 3 miles long with picturesque drops and views on both sides. The four top peaks of the ridge are capped with quite a different rock—a hard and whitish quartzite, formed about 600 million years ago during the Cambrian era. The structure one climbs today has been sculptured by glacial epochs of sliding ice and millions of years of rain and wind. We had lunch at the main peak, *Spidean a' Choire Léith*, 3,456 feet above the northeast end of Upper Loch Torridon, which from the final top blazed like a jewel in the afternoon sunlight. This day was so clear and

The author on the main top of Liathach, looking toward the final top above Loch Torridon. Fred Hoyle took this picture during the lunch break on the main peak.

Fred Hoyle atop the east face of Liathach. Beinn Eighe stands in the background.

bright that the horizon was distinct in all directions, and it seemed that the whole world was ours.

After we scrambled down the loose rock on the southwest end of the mountain, we faced the part we always liked least—the long hike back to the car. It was about 5 miles away, but fortunately a car came by and gave Fred a lift most of the way. Willy and I waited happily at a roadside stream with our feet in the water. There was always a daily ritual that had to be completed before earnest conversation could continue: Fred had to have tea, and Willy had to have a hot bath. I am a lover of ale, but Fred long ago showed me that the best thing after a day on the mountains was a cup of hot tea and scones with jam. Willy always happily postponed his bath for this portion of the ritual, so I assume that he also found it to be the best of all possible worlds. The subsequent hot bath, with dinner and wine following, was the main reason we always stayed at a rustic highlands hotel (Fred knows them all) rather than camping out. We never "roughed it"—it only *seemed* rough at times.

This particular hotel at Achnasheen was typical, built originally as a large private hunting lodge and run with a quiet and dignified friendliness. Before dinner it is necessary for Willy and I to pull out our copies of *Munro's Tables of the 3,000-Feet Mountains of Scotland*, a revised edition of the tables compiled by the late Sir Hugh T. Munro. Herein, with parenthetical markings, we annotate our accomplishments of the day. There are 276 separate mountains, or "Munros," in excess of 3,000 feet, and Fred had been up all of them but five. I have been up thirteen, and Willy, I think, nine—but no matter—one gets used to following in Fred's footsteps. The final part of the ritual is that Willy chooses the wine for dinner, a right he exercises with at least as much experience as Fred uses in choosing the mountains.

"Those chaps are simply determined to talk about an excess of faint radio sources rather than a deficiency of strong sources," Fred said as, without preamble, we again picked up the thread of an abandoned conversation. (In quoting him, I must here admit to quoting what I *think* I remember him saying, and I certainly cannot remember *verbatim.*) The subject was the number of sources of radio signals of various strengths that radio astronomers are measuring in the universe. Most of these cannot be identified with any visual object in an optical telescope, so their distances and physical natures are not well understood.

"Of course," Fred continued, "it's sensible to assume that *on the average* the weak sources are farther away than the strong ones, but the identified radio galaxies of known distances vary in power output over a range of 10,000 to 1, so the apparent strength of a source is not a good indicator of its distance. In fact, their whole argument is really based on the intrinsically strong sources, and there could just as well be a deficiency of 50 or so strong sources nearby as an excess of 10,000 faint ones farther away. You can't easily tell the difference between a universe that has fewer than average bright sources near the nearby systems of galaxies, and one that has a greater number than average about 10 to 100 times farther away."

The controversy was one that had rocked Cambridge and astronomy since the first radio surveys of the Cambridge radio astronomers at the Mullard Observatory antennae just a few miles out the Barton Road. They had made systematic errors in their first surveys and too quickly attacked the steady-state theory on the basis of inconclusive evidence. Now that the evidence is getting better, but only a little more conclusive, "those chaps" nonetheless stick by their opposition to the steady-state theory. Their arguments seem reasonable, but their groupish determination seems unwarranted. It was easy to detect signs of trouble in Cambridge.

To see how counting sources can show the geometry of the universe and its expansion, one should think first of a stationary Euclidean universe uniformly filled with radio sources of equal strngth. The apparent strength of each source would of course vary inversely with the square of its distance in a Euclidean universe, and the number of such sources within a given distance would vary with the enclosed volume, which is proportional to the cube of the distance. If the distance is eliminated from this relationship, one finds that in such a simple universe the square of the number of sources brighter than some specific brightness varies inversely with the cube of that brightness. Equivalently, the number itself varies inversely with the brightness to the 3/2 power. This result would also be true even if all sources had not the same absolute power because it is separately true for each different type of source —provided only that the density of each separate type of source be constant throughout the universe. One has only to count the number of sources brighter than some comparison value, regardless of their distances, and compare it with the 3/2 power of that brightness. It is a strictly geometrical argument, true only for Euclidean geometry. In retrospect, it would have been disappointing if the universe had been so simple—but it is not. Compared in this way with a bright group of sources, there seem to be about 2½ times too many sources 1/100th as bright; however, it is not clear whether the *average* number density of the more distant group of sources should be regarded as 2½ times more numerous than the *average* number density of the brighter group of sources, or whether, due to local irregularities in the universe, the brighter group of sources in the neighborhood of our galaxy is simply 2½ times less numerous than it is in an average location in the universe. The number 2½ is the discrepancy for Euclidean geometry. In any expanding universe, steady-state or evolving, the discrepancy in the number of bright and dim sources is even greater.

The attitude of the Cambridge radio astronomers to this may in part have been the result of an historical accident. They detected the few bright ones first, and later found more faint ones than they would have expected on that basis. Had they discovered the numerous faint sources first, they might have concluded, when they later found them, that there was a surprising deficiency of bright nearby sources.

"If we live in a portion of the universe having slightly lower density than average, so that about fifty bright sources are missing, the remainder of the counts, including about 100,000 sources only 1/10,000th as bright as the bright ones, looks as much like a steady-state or even Euclidean universe," Fred carried on.

"I guess that's if you assume the sources are constant in time," Willy added.

Fred agreed. "Those chaps claim that the discrepancy proves that the universe had more radio sources in the past than it has today, and that's why they see so many faint ones," he added. "Naturally, if the universe had more radio sources in the past than today, it is not in a steady state."

"Schmidt claims the quasar redshifts absolutely rule out the steady-state" I said, feebly trying to be devil's advocate by shifting the subject somewhat.

"That's right," Fred shot back, "if you believe those big redshifts are due to their being so far away in an expanding universe. Then I agree. But the quasar redshifts are so strange that I think they may be due instead to some new physics of the objects that we just don't understand. Some of them even have several different redshifts, and they can't *all* be due to the expansion of the universe. A strict steady-state having uniform density was already ruled out by Ryle's source counts. He is certainly right about that, but it seems to me that we more likely live in a grand steady-state that has large-scale fluctuations in space and time. These fluctuations may be the main subjects of astronomy."

So it goes. I don't believe I ever really helped Fred with his ideas. I just learned from him. But he seemed patiently willing to try out his thoughts, as if advocating them to us at least helped him keep them in order. I can't really say. It is always difficult, and even treacherous, to analyze the motives of a deep thinker—and one who doesn't even talk much at that.

"But does it go against cosmological principles to assume we're located within a region of the universe having lower than average density? And how big would such a hole have to be?" Willy wondered.

"Well, the hole would have to be a few percent of the Hubble distance," Fred responded, "and some people don't like that. But we see that the matter in the universe is clustered irregularly, and galaxies probably even occur in clusters of clusters. It doesn't violate the steady-state idea for matter to have large-scale clustering any more than it violates the idea of an expanding universe. Whether it's plausible or not depends on your system of values. After all, many of us find the ideas of a Big Bang and a universe expanding without end as pretty implausible too."

Dead tired, we went to bed by 10:00 P.M. Tomorrow we face another Munro, barring rain. Willy threatens, as always, to sleep till noon. Fred doesn't even blink. He himself may be up at 5:00 A.M. to get in a couple of hours of work on his ideas concerning conformal invariance before we meet for porridge and eggs.

Those were some days—never to be forgotten days.

As I finish these thoughts it is August 1972. Hoyle's Institute of Theoretical Astronomy just ended its 5-year existence on July 31, 1972. I shake my head in disbelief at the implausible events that conspired to terminate it. For six consecutive summers in Cambridge I found it to be the most exciting astrophysics center I have been in outside of Caltech. Largely due to Hoyle, IOTA was, during this period, a modern mecca. The pilgrimage of astronomers from Europe and the United States to its pastoral setting on the Madingley Road just west of the city center was a natural and almost inevitable result of Hoyle's unswerving intellectual vitality. Yet it seemed to me that the British and Cambridge scientific world admitted IOTA's cultural influence only grudgingly or took it for granted. So when the chips were down it seems to me that the Scientific Research Council of Great Britain was probably democratically unable to play favorites to the degree necessary to maintain this center of excellence, preferring instead to hope that it would continue to function as well under the aegis of Cambridge University as it had previously. The university, on the other hand, had not the leadership necessary for decisive action. Its committee rooms provided a stage upon which other professors could assert their right to allow Hoyle no more influence than they themselves had. So, amazingly, in a fiasco of vain struggling the nose was bitten off to spite the face. When the *Manchester Guardian* observed that, were it not for its medieval functioning, Cambridge University should have had some way of bending to Hoyle's special and even irreplaceable role, the new Cavendish Professor shot off a hot and defensive retort that the university committee on which he had served had acted as a collection of gentlemen. Hoyle, we are told, had as usual been a difficult character. No matter that he is the deepest intellect Britain has had in pure science since Dirac's formulations of relativistic quantum mechanics and the theory of antimatter, no matter that he had a masterfully conceived plan for the effective internationalization of British astronomy, a part of which already had brought Margaret and Geoffrey Burbidge back to England, no matter that Hoyle's Institute became an envied center of world thought—Hoyle had been a difficult character with the audacity to think that his accomplishments would earn him a stronger voice than the average in university affairs. Hoyle himself has resigned his Cambridge professorship, sold his house in Cambridge, and bought a home in the mountains of the Lake District, nearer to where his heart lies.

This particular case is perhaps not as interesting as the general phenomenon it manifests. I call the phenomenon *territoriality*, with intentional allusion to the anthropological concept. I have seen it in almost every scientist I have known and even in myself. One of its symptoms is that all scientists regard their own work as being of the top rank in importance, even though all know *intellectually* that some subjects *are* more important both philosophically and technically than others. If his subject itself does not warrant deep emotional involvement, the scientist often attaches that involvement to some detail that is sufficiently in doubt and sufficiently of interest to stimulate a little public discussion among contemporaries. When a scientist has contributed new information to a subject it becomes his territory. It is exhilarating if that territory has broad horizons of great interest, but it is sufficient that it allows him to lead a discussion with a dozen colleagues of similar interests. (Of course, we scientists do not speak of *territories;* we prefer instead the words *area* or *field,* which are at best psychologically slim disguises. Among the most common questions heard when one scientist is introduced to another is "What is your area (field)?")

The most obvious forum of territoriality is the scientific conference. I have known colleagues to exhaust themselves, damage their home life, and neglect their university duties in the compulsion to attend every important conference on a given subject—pulsars, say. The key moment in this drama is when the scientist is recognized at the conference and allowed to comment upon his particular expertise. These public utterances continually reestablish the scientist's territory. I cannot but think of Robert Ardrey's very popular and lucid, though professionally controversial, writings on the "territorial imperative" in animals. We are so like the simple birds who sing cheerfully to announce that the garden in which they perch is not inhabited by another male of the same species. A few scientists of especially grand verbal delivery always stimulate both the irritation and envy of their colleagues by their ability to confiscate abstract territory by timely, well-chosen, and controversial chirpings.

A less obvious forum is the university. Theoretically its structure is fluid so that it can adjust in such a way as to maximize its ability to transmit and create knowledge. In a practical sense this means that the sizes of and even existence of university departments can reflect the power and timeliness of their contribution. For example, my own Department of Space Science at Rice University was created by (then) President Kenneth Pitzer in 1963 to allow the possibility of a frontier intellectual effort in Houston to parallel the establishment of NASA there. One of the original members of our department, F. Curtis Michel, was also one of the original scientist-astronauts. But since university funds are limited, every such decision to participate strongly in some intellectual subject necessarily implies that some money will not be avilable for the study of some other subjects. The concept of territoriality makes it immediately clear that this can be an anxiety-ridden situation for those professors whose chirpings are relatively muted. A very similar competition exists for the national funds available for the support of scientific research. Territory pays!

I think these rather emotional concepts allow me my own understanding of what happened to Hoyle's Institute of Theoretical Astronomy. The personalities involved are strong and the minds are extraordinarily sharp. There is little effective outlet for such strong and aggressive personalities in science other than these irrational clashes of territoriality. People in other walks of life find more outlets in other kinds of empire building—earning money, acquiring land, giving instructions to the office personnel, etc. So, alas, the end came as surely as in a Greek tragedy.

The curious thing is that Hoyle himself was aware of the danger of such forces and still could not find ways to blunt their effects. He is seldom involved in public clashes over his imaginative and creative ideas. Hoyle may be little concerned over his own territory precisely because he has the security of knowing that so much original territory *is* his own. He is an outdoorsman through the rugged terrain of the frontier of knowledge, escalating each concept as surely as he scales each highland Munro. And territorial fences do not limit his contributions. He uses the ancient Scottish laws of public right-of-way to the tops.

CHAPTER XIII

ALL IN A DAY'S WORK

After breakfast in our home in Wiess College, the tree-lined walk across the Rice University campus is always a pleasure. It gives me a moment to order my thoughts and to enjoy the usual bright morning sunshine. The campus is a beautiful one, and I have always felt its small size and quiet atmosphere to be ideal for me. The large parklike grounds are an island of calm within the bustling aggressive boom city that is Houston. In many ways I must admit that I am a bit old-fashioned in my enjoyment of my simple life, but the complexity of the modern urban scene with its countless restrictions and possibilities holds little interest for me. I would gladly live in an earlier and simpler time except for one all-important fact: This same technological explosion has brought with it an associated explosion in our knowledge of the universe. I care very much about that.

Astronomy was for centuries synonymous with what the eye could see, albeit by rather impressive telescopes. The structure of the universe was inferred from the *light* that it emitted. Light is that special (to us) electromagnetic wave to which the human eye is sensitive, which means a narrow range of wavelengths—between 40 millionths and 60 millionths of a centimeter. Our eye developed that way because those are the same wavelengths that can penetrate the earth's atmosphere. The theory of evolution scores a smashing success here. Although the eye could easily have been constructed to see at longer wavelengths (the infrared) or at shorter wavelengths (the ultraviolet), such an eye would be looking at a dark world because those wavelengths are mostly absorbed in the earth's atmosphere before they can reach the ground. The human eye is in this regard an almost exact match to the transparency of the earth's atmosphere. At radio wavelengths near one meter in length the earth's atmosphere is also transparent, but an eye sensitive to radio wavelengths would be useless. Inanimate nature, however, knows no such bias toward light wavelengths. Laboratory physicists have become familiar with a continuous range of electromagnetic wavelengths ranging hundreds of millions of times longer and shorter than light waves. The external universe abounds with such waves, and the information they carry is correspondingly much greater than could ever be apprehended by the human eye. The application of modern technology to detect these "new astronomies" has opened the floodgates of new knowledge.

Another revolution in knowledge has been made possible by the modern electronic computer. In itself it knows nothing, but the computer has made possible lengthy calculations having a speed and accuracy that is literally astonishing. Calculations that would take years with pencil and paper can often be made in a matter of minutes. One can now use laws of physics to construct elaborate models of astronomical objects and check to see to what extent the behavior of the models is similar to the behavior of the objects one "sees" in the sky. The computer makes possible the simulation of nature.

As a result of all this we are confronted with such an increase of knowledge that one simply cannot keep up with it all. There was once a time when one active and gifted scientist could be familiar not only with most astronomical knowledge, but with practically all of physics as well. Those days are long gone. The curious scientist today bears facetious resemblance to a hungry man confronted by miles and miles of tables upon which are set the most exquisite and varied foods. He can eat some, taste several, smell and look at many, but there remain countless delicacies that to him must remain a mystery. Confronted with this, one can only hope to assimilate as much information as one can in the hope of finding those key ideas that may one day inspire a synthesis that explains much.

I don't always have such thoughts in mind as I complete the walk to the Space Science Laboratory, but on this day I did. Yesterday the November 15, 1972 issue of the *Astrophysical Journal* arrived in the afternoon mail. It is published twice a month by the University of

Chicago Press and contains the results of most of the important investigations in astrophysics carried out within the United States. When scientists have completed a piece of research they write a report of it and submit it to the journal for publication. Before it is published, however, the journal sends it to other contemporary scientists for evaluation. They judge whether it contains new results, obtained in valid ways, having sufficient importance to merit publication. Only then does it appear in print along with dozens of other articles in each issue. The November 15 issue on my desk contained thirty-eight such articles, and as I mounted the stairs toward my second-floor office I eagerly anticipated skimming it through for an initial appraisal of the relevance of its contents to me personally. I always do this because I never have time to read it all—or even a quarter of it. I knew I needed to spend some time working on my lecture to be given to a class after lunch, but I cannot resist the initial curiosity to see what the new issue contains. There would, I hoped, be time.

No sooner had I reached my door than I realized I had forgotten something. Richard Ward, a graduate student in the Department of

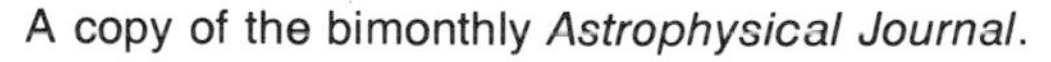
A copy of the bimonthly *Astrophysical Journal*.

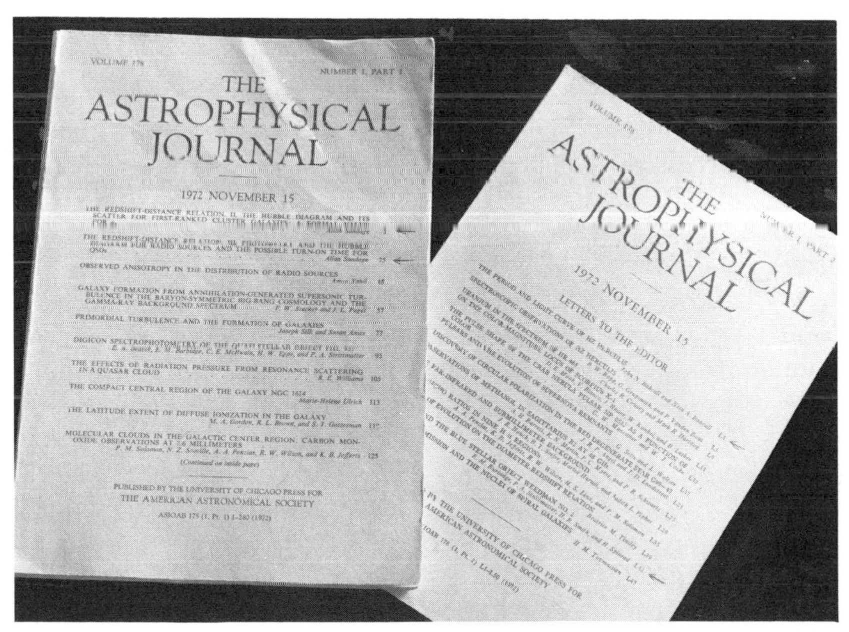

Space Physics and Astronomy, was seated outside my door waiting for our appointment. Richard obtained his Bachelor's degree in physics at Rice two years ago, and now he was beginning a research program that (we hoped) would lead to a Ph. D. In this effort I had agreed to help as research advisor because he is an extremely able student whose interests in cosmology parallel my own. Like most scientists, I believe very much in the value of the Ph. D. education. For the first time a student stops learning solely from books and lectures and begins instead to learn by trying to discover new knowledge. New knowledge! What a dizzying experience it is to discover things never before known to mankind, but that is exactly what Ph. D. candidates must accomplish within four years of graduate education. For them, there results increased opportunities in life besides giving confidence and satisfaction in having achieved a noble goal. For society, there results the value of increased knowledge and the value of having one more citizen with proven ability in solving a natural problem. This latter value persists even if the student turns to a wholly new occupation in life, for one never forgets the ability gained by a creative investigation that can meet all standards of contemporary criticism.

We had to discuss a problem in one of the new astronomies—gamma-ray astronomy. Gamma-rays are one manifestation of radioactivity of atomic nuclei. They are very short wavelength electromagnetic waves emitted from the atomic nucleus, usually after it has just emitted a beta ray in another form of nuclear activity. Three years before I had formulated a theory that predicted that the universe should be full of such gamma-rays in a very specific range of energies. These gamma-rays would have traveled over most of the distance of the knowable universe because they would have mostly been produced very early in the history of the universe when the largest portion of the atomic nuclei were created. I had argued that common iron atoms were actually created as atoms of radioactive nickel in explosions of stars. The radioactive decay of these nickel nuclei in space would have created a quite predictable spectrum of gamma-rays that would be just now reaching us after their immense journey through the universe. The theory is hard to check because the gamma-rays cannot be observed in the earth's atmosphere, which they cannot penetrate.

The thing now exciting us was that the gamma-ray detector carried aboard the *Apollo 15* spacecraft to the moon brought back evidence of gamma-rays in space that could be just these I have expected. If so, we may learn quite a lot about the history of the universe from them—but first many problems must be solved. Richard will be trying to construct a computer program that can compare the *Apollo 15* measurements with the detailed expectations for several different plausible histories for the universe. After half an hour of conversation about it, he went away with several ideas (mostly his) about how this might be best accomplished.

Richard's visit also reminded me that I have a potential problem with the financial support for this research. As he left I decided I should telephone the National Science Foundation in Washington. Universities do not have adequate funds to support by themselves the costs of these research efforts—the tuition and modest stipends of the research students, the fees for running the computers, the equipment for the laboratories, and the costs of having the new findings published. Society itself must do this or it cannot be done. The National Science Foundation was created largely for this purpose. Scientists submit proposals for research support to the Foundation describing what they wish to investigate and how much it would cost. After a critical, confidential review by scientific peers, the Foundation selects those projects that it can support with the limited funds appropriated to it by Congress. The competition is spirited, because the Foundation is given just enough money to keep scientific research active, but by no means enough to support all or even most of the worthy proposals it receives. I cannot decide impartially if society supports these attempts at a level that is consistent with the benefits to society of scientific education. The annual budget of the Foundation is about $500 million, which is about $2.50 per person in the United States. Congress and the people must decide how appropriate that amount is. The last few years have seen the money spent on university research and education reduced unless that research is judged to have some immediate application to a national need. Like all my colleagues I am anxious about next year's funds for my research. When I got through to the Foundation, I heard that my proposal for continued support had gotten good evaluations, and they hoped to be able to support most of the costs. That was good news. I went to get a cup of coffee from our community pot.

It was later in the morning than I had hoped as I picked up the *Astrophysical Journal* to see what was in it:

(1) Two Princeton scientists measured the brightness of the star HZ Herculis as a function of time to see if its brightness is constant. Their reason for selecting this star from the millions in the nearby sky was that an earth-orbiting satellite called "UHURU" had recently reported a strong source of x-rays in the sky very near the position of the star HZ Herculis. UHURU was launched by the United States specifically to look for astronomical sources of x-rays, and it found that this source in Hercules varied in intensity. The period of its x-ray variations is 1.70 days. This new paper shows that the light from the star has the same period; therefore, the x-ray source and the star are part of the same object. The most plausible model is an x-ray emitting neutron star in orbit about an ordinary star with a 1.70-day period. This could be the first conclusive observation of neutron star—matter so dense that one cubic centimeter would weigh a billion tons on earth! I wonder how it happens that an ordinary blue star can be in such a close orbit with a neutron star. The only thing I can suppose is that the neutron star is all that is left of an ordinary star that exploded.

(2) The next paper reports that three University of Texas astronomers identified the optical spectra of some common atoms in the same star.

(3) Two spectroscopists at Michigan have identified the element uranium in a star. That's a surprise to me because I didn't think a star could produce uranium without exploding. The uranium concentration in this star is about a million times greater than the uranium concentration in the earth! The rare earth elements are also very overabundant in the same star.

(4) Astronomers at the new United States telescope in Chile have studied the optical color of another recently discovered source of x-rays in Scorpius. It's rather blue, but its color varies.

(5) University of Arizona astronomers have been looking again at the color of the visible pulsar in the Crab Nebula. The Chinese saw this star explode in 1054 A.D., and that must be when the pulsar —probably another neutron star—was born. It is the only pulsar one can actually see blinking, in this case thirty times per second! The

curious thing they find is that the shape of the repeating pulse is different in different colors.

(6) The next paper presents theoretical argument that suggests that the strength of the radio emission from a newly formed pulsar should not be as great as many expected. If their argument is correct it explains why we do not detect any very bright, very young radio sources.

(7) The light coming from the red "white dwarf" G99-47 is not randomly oriented. The electric vectors of the light waves are circularly polarized. I'm not sure what the significance of that is, but white dwarfs themselves are fascinating. With densities of tons per cubic centimeter, the atomic nuclei are still not squeezed together. The whole crushing gravity is resisted by an esoteric principle of theoretical physics called the "exclusion principle."

(8) MIT radio astronomers have discovered methanol (CH_3OH) in an interstellar cloud about a million times more massive than the sun. It sounds like Hoyle's "Black Cloud." It also contains formaldehyde, ammonia, carbonyl sulfide, carbon monoxide, and methyl cyanide. Maybe the chemical basis for proteins existed even before the solar system formed. These scientists traveled from Massachusetts to the National Radio Observatory at Kitt Peak to make their observations. That is the usual procedure these days. The National Science Foundation pays the relatively small travel costs to use the big national scientific facilities.

(9) The Cornell group launched another Aerobee 170 rocket from the White Sands Missile Range to make another check on the temperature of the universe. They confirmed again that the universe is like the inside of a very cold oven—to be exact, 3° absolute. That's −270°C or −486°F. However, they did not detect the anomalously high flux of electromagnetic waves with wavelengths near one millimeter that they reported some time ago. They conclude that the universe cannot have converted a significant fraction of its mass into electromagnetic waves.

(10) Bell Telephone scientists detected the rare isotopes of carbon and oxygen in regions of ionized hydrogen. Their rarity there does not appear to differ much from their rarity on Earth, however.

(11) The next paper argues that because galaxies do not maintain constant brightness as they age, one must be very careful in comparing their observable sizes with their distances. The problem is that a

photograph does not measure the actual angular size of a galaxy; it shows only the brightest central portions that are brighter than the background light of the night sky.

(12) University of California astronomers have found a small blue object resembling a quasi-stellar object (QSO) very near two apparently disturbed galaxies. The two galaxies are not very distant, so if the blue object is a QSO related to the galaxies, it is also not very distant. They are suggesting that one might find here further evidence that QSOs are not at the great distances one might have supposed from their large redshifts. Arguments about this sort of thing have raged for the past five years, and we still don't have a conclusive answer. Are quasars the most distant visible objects in the universe or not? Obviously someone must see if this blue object is a QSO or not, and, if it is, what is its redshift? One of these authors is Margaret Burbidge, who in 1972 became the first woman ever appointed director of the Royal Greenwich Observatory.

(13) A Russian scientist, visiting the United States at the National Radio Observatory in Green Bank, West Virginia, reports one result of their new system of classifying galaxies. It is apparent that strong radio emission tends to come from galaxies that have very bright central regions called nuclei. He concludes that bright galactic nuclei are not properties of very stable galaxies but rather of galaxies that are undergoing active evolutionary changes. This problem of the evolution of different galactic types has been a prominent one ever since the various types have been noticed and described. The "new astronomies" are adding a lot of fuel to those fires.

(14) Allan Sandage has published the latest results of his continuing investigation of the relationship between the apparent brightnesses of galaxies and the size of the redshifts of their optical spectra. Hubble demonstrated the relationship in a paper published 44 years ago, and Sandage, as his spiritual successor, has spent the past two decades at the same Mt. Wilson Observatory refining many aspects of the correlation. It is the basic program in observational cosmology. One can only admire Sandage for the lively intellect and fierce motivation that have allowed him to endure the continual frustrations of that quest. In this latest paper he uses data on apparent brightness of the brightest galaxy in each of eighty-four clusters of galaxies. He finds the smallest value yet of the Hubble parameter measuring the velocity of recession of the distant galaxies; 16 kilometers per second faster for each million light years of

extra distance, 1/10th of Hubble's original value. He also finds a suggestion of deceleration in the expansion of the universe, roughly $q_0 = +1$. If these numbers rightly describe a Big Bang cosmology of the Friedmann type, then the universe is about 11 billion years old—slightly more than twice the age of the earth. The problems associated with the interpretation of q_0, however, remind me that I want to start a new student investigating the effect of aging on the brightness of the galaxies. It is also interesting that the brightnesses are measured by photoelectric photon counters built by Astroelectronics Laboratory, rather than by old-fashioned photographic film. One interesting thing Sandage finds is that the accuracy of Hubble's Law is limited primarily by the fact that the brightest galaxies in different clusters of galaxies do not have the same absolute brightness. Thus their relative brightnesses do not give an accurate indication of their relative distances. We badly need a theory of the evolution of galaxies that can tell us how bright a galaxy is by other features like its shape and color. That sounds like a difficult task, but it has worked for stars.

(15) The next paper is also by Sandage. It is a study of galaxies that are strong sources of radio emission and of quasars (or QSOs) as well. The radio galaxies show the same Hubble's Law as the brightest galaxies in clusters, but they are slightly fainter. Apparently a galaxy must be more massive than 10^{12} (a million million) suns to be a strong radio source. That's about ten times more massive than our own Milky Way Galaxy. He argues that quasars are brighter than radio galaxies but that their brightnesses differ greatly. He also argues that the reason no photons from quasars have their wavelengths increased by more than three times their usual wavelength is that no quasars were born until the universe was more than one billion years old. This point of view is in marked contrast to a previous paper in this issue seeking to see if there is some physical association between quasars and adjacent galaxies that are known to not be at great distances.

(16) In Princeton a visiting Israeli scientist has studied the number of radio sources as a function of their brightness and found that the northern hemisphere differs from the southern hemisphere. If that holds up it either means that the distant universe is not the same in all directions or that a large number of unidentified sources in the northern hemisphere are weak sources fairly nearby. This paper will really set the cosmological tongues wagging.

(17) Two theorists consider how galaxies could form early in a Big Bang universe containing equal amounts of matter and antimatter. They argue that (1) galaxy formation was caused by turbulence in the young universe, and this turbulence was driven by annihilation radiation at the boundaries between clumps of matter and antimatter; (2) the early universe was an emulsion of regions containing nearly pure matter and nearly pure antimatter; (3) the flux of cosmic gamma radiation around one MeV of energy is due to the early matter-antimatter annihilation, with the roughly seventy MeV gamma-rays produced in that way redshifted by a factor of about seventy in the subsequent expansion of the universe. I should study this one because he will attempt to explain by this scheme the same gamma-rays that Richard and I will try to explain by cosmic radioactivity. We'll probably both be wrong! I will be going to NASA Goddard Space Flight Center in April for a meeting on cosmic gamma-rays, and perhaps I can discuss it with them there.

(18) Two Berkeley scientists also consider the effects of turbulence on the formation of galaxies. Their picture suggests that the clusters of galaxies form after the galaxies themselves and provides a plausible account of the observed angular momentum of spiral galaxies.

(19) California astronomers have used a new electronic device, a digital image tube, to measure the absorption due to hydrogen in the quasi-stellar object (QSO) named "PHL957." This "digicon" allows the intensity of the light to be quickly measured as a function of wavelength, and it does so by actually counting photons rather than requiring, as does film, ample energy for "exposure." They conclude that the absorption lines arise from the QSO itself rather than from absorbing matter in intergalactic space. The redshift is large because the observed hydrogen light has a wavelength 2.3 times longer than it was when it was emitted.

(20) A University of Arizona astronomer considers whether the pressure of the bright light around a QSO might be great enough to accelerate shells of material to velocities near the velocity of light. If so, such a model could explain why many QSOs show atomic absorption lines whose wavelengths have been redshifted by considerably smaller factors than have the wavelengths of the atomic emission lines. (It could work, I guess, but the whole situation sounds very implausible to me.)

(21) Photographs of the spiral galaxy NGC1614 show it as a compact galaxy with a patchy core, spiral plumes, and a long blue jet suggesting that it has suffered a major explosion or eruption of some kind. Spectrographic studies from the University of Texas now show that about one billion times the mass of the sun is concentrated in the central region of about 1,500 light years (less than 1 percent of the area of the spiral galaxy) and, furthermore, an ionized gas is flowing out of the core with a velocity of 450 kilometers per second. This galaxy is also a very bright infrared object in the sky. It is certainly doing something dramatic and important.

(22) At the low gas density in interstellar space there exist oversized hydrogen atoms some tens of thousands of times larger than ordinary unexcited hydrogen atoms. Instead of light waves they emit radio waves. National Radio Observatory astronomers have observed the shape of our galaxy in 18-centimeter-wavelength radio waves from such atoms, and they find that the hydrogen gas in our galaxy is as flat as a pancake. Its distribution is only 200 light years thick, in contrast to the near 100,000 light years from one edge of our galaxy to the other. The gas is held in that thin pancake by its own weight. The gravitational attraction to the massive stars in the galactic disk provides the source of the weight of the gas.

(23) The carbon monoxide molecule (pollution!) emits an electromagnetic wave a quarter of a centimeter in wavelength. It has been observed coming from the center of our galaxy.

(24) Livermore Laboratory scientists use a statistical argument to determine the absolute luminosities and distances of twenty x-ray sources in the sky. Only the directions and apparent x-ray brightnesses of these sources are known with accuracy. We need some additional arguments like these to determine how far away they are. They find that the strong x-ray sources lie in a loose cluster about the center of our galaxy—30,000 light years away.

(25, 26, 27) The motion of a shock wave in the interstellar medium is considered in the next theoretical paper. Calculations of the high density gas after it has cooled subsequent to the passage of the shock wave is compared to the well-known Cygnus Loop. The same University of Wisconsin astrophysicist follows with another paper on the cooling of the gas following the explosion of a star (the "supernova remnant"), and a third providing a detailed model of the Cygnus Loop.

(28) When light scatters from a free electron, the scattered photon usually has a slightly longer wavelength than before it was scattered. In the next paper the effect of repeated scatterings on the shape of the spectral emission lines of atoms is calculated. It is uncertain if this is applicable to actual astronomical objects. Richard and I should take this into account in our gamma-ray study.

(29) The next paper lists the location in the sky of twenty-three faint objects emitting spectral lines. Their excitation temperatures are unusually low and the authors speculate that they might be planetary nebulae in the process of formation. The planetary nebulae, beautifully colored spherical shells of glowing gas surrounding a central star, have long been an observing favorite of amateur astronomers. If these twenty-three objects are planetary nebulae caught in the act of formation, their study sure might reveal how it happened. I have been assuming they are formed when red giants turn into white dwarfs, but the sequence of events has been more like a gut feeling than a theory.

(30) The next paper is the fifth in a sequence describing ultraviolet observations of stars from a U.S. satellite launched solely for astronomical purposes—the so-called Orbiting Astronomical Observatory. Because of the absorption by ozone in the earth's atmosphere, hard ultraviolet rays do not reach the ground. From the satellite it is possible to electronically measure the brightness of blue stars at several different ultraviolet wavelengths and thereby to determine the ultraviolet brightness distribution of stars. This paper concentrates on blue stars that have abnormally weak helium lines—an important question for cosmology considering the current mystery concerning the origin of helium in the universe. Their evidence suggests that the temperature structure in the atmosphere of the helium-weak stars is not different from normal stars of the same color. The basic puzzle about them remains.

(31) Electronic observations of the brightness of stars in the well-known galactic cluster, Hyades, have revealed that one of them is a flare star. On December 5, 1970, the brightness of that star viewed with an ultraviolet filter increased by a factor of fifteen during a 10-minute interval. In a half hour it was back to normal. I wonder what causes such a sudden change in the brightness of an ordinary star. If the sun did that, we'd all get pretty good sunburns.

(32) Two Australian astrophysicists have constructed numerical

models of stars that they suspect may cause the planetary nebulae. A sequence of 440 models was calculated by the computer showing what the star does as time passes—in this case about 10,000 years of "star time" elapses in a few hours of "computer time." The authors assume that nuclear reactions in the inner 25 percent of the star's mass have converted hydrogen and helium into carbon and oxygen, while the remaining 75 percent is pure helium. The nuclear power comes from a spherical shell in the helium region where helium is fusing into carbon, and they find that when that thermonuclear-fusion shell approaches the surface of the star, a thermal instability occurs. The power output rises to about 10,000 times the power of the sun, at which point it appears likely that the pressure due to the intense radiation pushes off the outer shell of the star, leaving the central portion to cool off to a white dwarf. Whew! The whole thing should very much resemble the observed formation of planetary nebulae.

(33) The next paper is a very extensive theoretical treatment of the solar wind—a 400 km/sec flow of very low-density ionized gas outward from the sun. This gas has about ten protons and electrons per cubic centimeter as it rushes past the earth. That's an indescribably better vacuum than one can construct on Earth, but because of its high speed this flowing plasma can cause numerous puzzling phenomena in the earth's upper atmosphere—most noticeably providing particles for the Van Allen radiation belts and the aurorae or northern lights. The main point of this paper seems to be a detailed calculation of the temperature of both the electrons and the protons. He presents results for the directional distribution of proton velocities at the orbits of the outer planets.

(34) During solar flares the erupting region emits electromagnetic waves with wavelengths of several centimeters. This radiation comes in bursts. Stanford University scientists present calculations to substantiate their view that this radiation is caused by the spiral motion of fast electrons in a magnetic field. The fast electrons are, in their view, accelerated in the solar flare. It is certainly a well-established fact that fast moving electrons in a magnetic field do radiate. I remember even seeing visible light from this process in the electron accelerator at Caltech. It is called "synchrotron radiation." This kind of analysis should really help in understanding the physical conditions in solar flares.

(35) Another paper about the sun attempts to explain the patchy appearance of its surface brightness. This patchiness shows clearly in photographs of the sun taken with a special telescope that was lifted to the top of the earth's atmosphere beneath a helium-filled balloon. The authors interpret the brightness fluctuations as being temperature fluctuations produced by turbulent convective motions (boiling!) of the solar surface.

(36) The next paper reports the discovery that sunlight reflected from the planet Saturn is circularly polarized. The meaning of this discovery is unclear.

(37) University of Colorado astrophysicists have gathered more information about how the earth's atmosphere absorbs electromagnetic waves having wavelengths near 1 millimeter. These waves are so strongly absorbed that they can hardly be detected at ground level, so they traveled to Baja, California, where an observatory of the University of Mexico is located at an altitude of 9,250 feet. There they observed the millimeter brightness of the sun at different times of day when the slant paths of the sun's rays through the earth's atmosphere are of differing lengths. They found that the absorption was anomalous—in the sense that it is greater than can be accounted for by water vapor and known minor atmospheric constituents. It is really curious to me that even today the absorbing properties of the earth's atmosphere remain puzzling.

(38) In the final paper NASA scientists report laboratory measurements of the x-ray lines emitted by highly ionized aluminum atoms. Detailed knowledge of such wavelengths find many applications in the modern astronomies. X-ray spectroscopy of the sun from satellites reveals countless lines that have never been identified in terrestrial laboratories.

Well, that's it. This much new information appears every two weeks when a new issue comes. I will try to read several of these papers more thoroughly, but just now I note that it's ten minutes past noon. I want to go home to have lunch with Annette. Then I will try to do more preparation for my afternoon's class. It's all in a day's work.

CHAPTER XIV

YOU CAN'T RUN IT BACKWARDS

My father is an enthusiastic photographer. As I grew up, I can remember a long line of equipment—cameras, projectors, enlargers, seemingly huge bottles of developer, and a kitchen turned darkroom at night by black paper and tape. A lot of childhood fun came from this, and the most fun of all was the home movies. My brother and sisters could be seen riding tricycles or bicycles, or perhaps running through the lawn sprinkler or diving into the University Park public swimming pool. We also had animated 8 mm cartoons, without sound of course. We children provided that, especially when the film ran at unnatural speed or in reverse. The reverse films tickled me to death! I remember rolling on the floor doubled up in laughter at some of the bizarre events. Nor did my subsequent education in physics reduce my amusement. Quite to the contrary, the physicist thinks often about the direction of time. "Can I run it backwards?" It is a basic question about fundamental interactions and about the universe. "The flow of time," as Newton called it, is one of the commonplace concepts in physics and in life, but it is still used without a secure understanding of the true nature of time and why it seems to flow in one direction. Have you ever wondered, as I did, why the film is so funny in reverse? And have you ever wondered, as Ponce de León did, if a man can grow younger?

Of all the mysteries surrounding life, the one that teases me the most is that such well-ordered structures as ourselves should emerge in a universe dominated by a propensity toward chaos. This observation has been offered throughout the span of man's existence as evidence for creation by a deity. Even in life, however, the inexorable advance of natural events toward disorder and chaos produces sure death. Poets and modern biologists alike have observed that the slow process of death begins immediately after birth as the initially well-ordered organism suffers gradual mutation of its microscopic structure. We replace all our cells many times over in a lifetime. The secret of life seems to reside in its ability to pass on information via a genetic code and replication system so that the species continually renew themselves—at least until conditions become so unfavorable for the species that it becomes extinct. Yet how marvelous the whole procedure is! Modern science has not removed that mystery, and in many respects the mystery has grown as knowledge has expanded. People still argue with the helplessness of children the question of life on other planets in the universe and its similarity or dissimilarity to our own. That no conclusive answer has yet come forth (although we now know the moon is lifeless, and we should easily know the answer for Mars in our lifetimes) is testimony to our basic lack of understanding of the miracle of ourselves.

The study of physics has provided a fundamental law that lies at the basis of much of this puzzle. I now remember the law well because I stumbled over it during my Ph. D. oral qualifying examinations! This law, which is as important in the realm of physics as the existence of the four basic forces and the expansion of the universe, asserts that natural events proceed in the direction of things becoming more disorderly and more chaotic. It has never been found to be violated, although sometimes we must think hard to see how the law is operating. To understand this law, one must consider how it is that man was led to it (in a *scientific* rather than a poetic sense). One must ask himself how *order* and *disorder* are to be measured, because the physicist chooses to deal with measurable (i.e., numerical) concepts. If you care to appreciate this quandary, digress with me briefly into this realm of human knowledge.

It has always been regarded as a beautiful and true symmetry that the laws of physics have seemed not to depend microscopically on the direction of time. For that reason, the laws of physics are presently formulated with the attractive feature that if an event obeys the laws, the event reversed also obeys the laws. Imagine a film of the collision between two billiard balls on a billiard table. The balls carom from each other with directions and speeds determined by their initial speeds and how nearly straight-on was their collision.

If the film is now shown backwards it will also present a sensible collision between the two balls. Nobody would laugh. Indeed, each glancing collision of the two balls has a perfectly possible counterpart in the backward running film, and it is satisfying, therefore, that the mathematical laws describing this collision do not depend upon the direction of time. The forward and backward running films differ only in the sequence of the initial and final conditions.

The same thing is not true of more complicated phenomena. Show any film in reverse of traffic at a busy intersection and the audience will immediately break into laughter. Everyone appreciates such hilarious action, and yet what is it that makes it so funny? A little thought shows that the film in reverse is not really physically impossible—it's just so *improbable* that it appears ridiculous. It is improbability, not impossibility, that distinguishes the direction of time and gives rise to the famous "Second Law" of statistical physics. It asserts that the more complex the phenomenon, the more probable it is that the system can only pass from order into disorder and not the reverse. Consider again the billiard table where the first shot consists of one ball striking ten balls symmetrically placed at rest in the form of a pyramid and dispersing them in all directions. When that film is seen in reverse we see eleven balls colliding at a common area in such a way that ten of them stop in a symmetric arrangement, and one ball speeds off carrying all of the net momentum of the original eleven. It is not an impossible event, and if all eleven balls could be given initially exactly the correct speeds and directions, it could actually happen; nonetheless, the event looks funny because it is fantastically improbable. The reversed film shows us the unlikely event of creating order from apparent disorder. The improbability of the reversed event comes about not because the laws of physics are violated but because the increased number of balls has made the reversed event so improbable. How much more improbable would be a reversed film of an even larger number of particles. Considering that even the smallest visible chunk of matter or volume of gas has billions of billions of molecules, we see that a disordering process would be so improbable when viewed in reverse that it would be essentially impossible. An example can easily be imagined by dropping ink into a glass of water. The random collisions of the molecules soon distribute the ink molecules uniformly through the water, creating a chaotic dispersal of the initially well-ordered ink droplet. If we simply watched the individual molecular collisions of this film in reverse, we would sense that each collision *is* possible in reverse. What would defy the odds would be for all of the billions upon billions of collisions to act together in just such a way that all of the initially dispersed ink molecules come back together into a well-ordered droplet of ink. No wonder then that we never see a man grow younger!

We measure this order and disorder by a physical quantity called the *entropy* of the system. The entropy increases as the disorder increases, and the disorder is measured by the number of different ways something can happen. The greater the number of ways something can happen, the more likely it is to happen, whereas things that can occur only under special circumstances are unlikely. I finally understood how this is measured quantitatively by recalling a childhood fascination with the distribution between heads and tails as coins are placed upon a tabletop. If there is only one coin, it may be either a head or a tail with equal probability. With two coins there is only one way to have two heads—namely, that both coins be heads—whereas there are two ways to have one of each—namely, the first as a head and the second as a tail or the first as a tail and the second as a head. Clearly one head and one tail is twice as probable as two heads. See how much more quickly the numbers grow when we increase the number of coins to four. Let us name the coins A, B, C, and D. Here is the list of the possibilities.

4 heads, 0 tails: one way possible
heads (A, B, C, and D)
3 heads, 1 tail: four ways possible
heads (A, B, C); tails (D)
heads (A, B, D); tails (C)
heads (A, C, D); tails (B)
heads (B, C, D); tails (A)
2 heads, 2 tails: six ways possible
heads (A, B); tails (C, D)
heads (A, C); tails (B, D)
heads (A, D); tails (B, C)
heads (B, C); tails (A, D)
heads (B, D); tails (A, C)
heads (C, D); tails (A, B)
3 tails, 1 head: four ways possible
4 tails, 0 heads: one way possible

In the random falling of four coins there are these sixteen possible results, of which the most probable is two heads and two tails, which can happen in six different ways. One easily sees that if the number of coins is made large, say, one billion, the number of ways to have approximately equal numbers of heads and tails is much greater than the number of ways to have, say, two times as many heads as tails. And the number of ways to have each be heads is still just one! The physicist defines the entropy as the logarithm of the number of ways the system can happen, so the entropy of the system containing all heads is small and the entropy of a system containing nearly equal numbers is large. The connection with *order* in general is obvious. All-heads is a highly ordered system, whereas roughly equal numbers of each is not. In just such ways as this the disorder of a system is quantitatively measured by its entropy.

These ideas allow me to imagine more clearly the meaning of the Second Law: Spontaneous changes in the total system never proceed in the direction of decreasing entropy. That is, *an isolated system never imposes order on itself.* Imagine that the tabletop is made to vibrate just strongly enough that coins jiggle about and occasionally are flipped over. A film of the motion of these coins would show that the initial state of the all-heads is gradually changed to the state of increased entropy for which the numbers are nearly equal. The film running in reverse would appear ludicrous—a large number of randomly bouncing coins would slowly be seen to be changing to a preponderance of heads until (miraculously) each coin is heads. No man would believe it. It would represent a case of nature spontaneously decreasing its entropy. So everyone believes the Second Law of statistical physics—even if he doesn't know its name.

Sometimes the Second Law is expressed in terms of heat flow, in which case it is called the Second Law of thermodynamics: namely, heat will not spontaneously flow from a cold body to a hot body. The reasoning is the same: Although it is *possible* for heat to flow from cold to hot (i.e., running the film in reverse), it is so fantastically improbable compared to the normal flow from hot to cold that it is virtually impossible. If we could see the atoms of any substance, we would see them bouncing around and colliding with their neighbors with an energy that increases as their temperature increases. The total amount of energy is conserved in these atomic collisions, but it may be transferred from one particle (usually the faster moving one) to another (usually the slower moving one). When a hot body is in contact with a cold one the more energetically vibrating molecules of the hot body give energy to the more slowly moving molecules of the cold body and a transfer of heat energy from the hot body to the cold body is affected. For the collisions to occur in such a way that most of the slower (colder) molecules rebound even more slowly while the faster moving (hotter) molecules rebound with even more speed is possible in principle, but the odds are so against it that it may be regarded as virtually impossible. For the cold body to become colder and the hot one hotter requires the system to spontaneously reduce its own disorder by concentrating the available energy into one of the two bodies. This violates the Second Law.

One hot summer Texas day when I was thirteen, my father bought a miraculous machine in which heat flows from a colder body to a hotter one—an air conditioner. It removes heat (molecular speed) from the air molecules inside the cool room and deposits the heat in the hotter outside air. So impressive a miracle was this that my subsequent understanding of the Second Law was almost torpedoed in advance. Does it really violate the Second Law? Not at all, because it does not happen spontaneously. To move the heat energy from cold to hot requires work from another machine, and that machine must be powered ultimately by an even larger amount of heat flowing in the normal direction. For example, the air conditioner is usually driven with electrical power which is in turn generated by steam turbines. They get their power from the burning of fuel, which makes steam, and from the hot steam molecules attempting to expand into the colder surroundings. This process cannot be completely efficient, so a lot of thermal pollution results at the source. When the whole system is considered, one sees that even in an air conditioner we have a net transfer of energy from a hot source (ordered energy) to the cold surroundings (disordered energy). And so it is generally that the creation of order from disorder does not happen spontaneously but requires some machine receiving energy from a hot source and returning it to a cold source.

These thoughts suggest the machine that keeps life running on the earth. It is, of course, the sun. It is a property of vegetation that the chlorophyl readily absorbs the high-temperature energy from the sun. This energy is utilized in the replication process and then reradiated into cold space as infrared radiation. The earth radiates just as much energy as it receives, but it receives it from a high-temperature source (sunlight) and reradiates it as low-temperature energy (infrared) into the cold emptiness of space. The entropy increase (disorder) associated with this overall process is large and allows a smaller amount of entropy decrease (order) to emerge as life. In that sense life is like the air conditioner, although we may not be fond of the comparison. In both cases the restricted event that at first seems contrary to the Second Law is really part of a much larger machine that does not violate it. There's poetry in that. Life is not isolated from the universe. We require the blessings of high-temperature energy and a cold reservoir (the universe) into which the same amount of energy is returned as low temperature radiation. One might have thought that all the sun does is keep us warm, but the situation is more dramatic. If the earth were placed inside a large incubator whose walls were at the average temperature of the earth's atmosphere, the earth would be just as warm as it now is, but life would not survive. We would be radiating energy at the same temperature we receive it, and there would thus be no overall entropy increase that would allow a local entropy decrease (life) to occur. Slowly plant life would wilt and die and the earth would become a warm randomly disordered grave. We need to dump our thermal pollution into cold space and receive new supplies of hot sunlight.

Unfortunately, even this understanding of how it is that the occurrence of life does not violate the Second Law does not allow us to predict that life *will* happen. All one can say is that the existence of life does not violate our laws. But we still do not understand that given these circumstances *life must occur*. It may still be a fantastic accident that might not have occurred on many other planets with equally favorable conditions. We just do not know. There is no physical law (though it has been suggested) that asserts that *everything that can happen does happen!* The issue here is difficult and will not be soon resolved, for it is not just the chance occurrence of complicated molecules that we seek, but a driving principle to explain why a physical system would produce an order capable of passing on information to new replicas of itself and how that information transfer ever got to be so sophisticated as in common man, where we find nature consciously studying itself.

The astronomical setting of life elsewhere in the universe is more complicated than might at first have been thought. It depends on the temperature of the central star of the planetary system and on the radiation frequencies that easily penetrate the atmosphere of the planet. These things will differ from planet to planet, and so also may the life that emerges there. The evolution of species here on Earth has invented more than 100 million different species of plant and animal, and 98 percent of them are now extinct. The normal expectation of any species, if it can expect at all, is that it will become extinct. That is the statistical likelihood for us too. What ever happened to Neanderthal man? He lived when we emerged, and it is not difficult to guess who extincted him. Who extincts the salmon, the whooping crane, the wolf, the whale, the eagle? Can we guess? Extinction occurs not solely because of active predators—for the mouse is not extinct—but also because domination by one makes the environment unfavorable for many. The usual pattern is that some better creature will come along and make us extinct. Man has, in addition, the terrible capability of making himself extinct. If we are to avoid either fate it will surely be because we are the first organism on earth to understand and control evolutionary processes. It is now almost within our power to cause our species to develop along whatever lines seem best. Have we the judgment for such a decision?

There is something frightening and inevitable here, but it is not to be imagined as "nature, red in tooth and claw," pouncing, killing, and devouring. The thing too horrible to mention is more like the cold certainty of statistical physics. When water on the stove becomes hotter than 100° C, it turns to steam; and unless the steam is carefully captured and held in place the film will not run backwards. When helium is cooled below −271° C, it becomes a superfluid. A metal cooled below a critical temperature loses its electrical resistance. These things are called *phase transitions*, and they represent collective properties of a species. The collective properties are not exhibited by individual members of the species, but appear as inevitable statistical results when the number within a species becomes large and when the environment becomes critical. The question of survival of each species of life bears resemblance to a phase transition between the state of abundant life and the state of extinction. How long has the whale lived? Very long. How quickly does the whale now become extinct? Perhaps very quickly. Is there anything we can do now to help the whale survive? According to some experts, it's very doubtful. A mere 20 years ago—yes—but now it seems too late. The few remaining cannot even find one another within the huge oceans for mating. How long has man been the dominant species of earth? Probably only about 10,000 years. Dominance became possible after the widespread development of agriculture, an invention so important that it critically altered the balance of power. It allowed part of the population to engage in occupations other than the pursuit of food. It made possible the first great civilizations. I only wonder what it means that 10,000 years is less than one part in 100,000 of the total span of life on earth.

That nature moves statistically in one direction only, must have consequences just as grave for the universe as for the gambling table. The universe too seems to move in one direction only —expansion—and it is hard to know if it could run the other way. The hot stars radiate hot light into cold space, thereby increasing the entropy of the universe. The power for this comes from spontaneous thermonuclear reactions converting light elements into heavier ones within the stars. The great energy release of each reaction is quickly distributed in a statistical way among the particles there. I cannot imagine this film in reverse in a plausible way—light from cold space being absorbed by the surfaces of stars, and diffusing inward as heat toward higher and higher temperatures, whereafter the statistical heat is miraculously appearing as big chunks capable of turning the heavier elements back into lighter ones. No—the universe we see can only be running in one direction. But why then hasn't the activity within the universe run down? Surely it cannot run forever as I envision it. The great power output of condensed objects like stars, galaxies, quasars, pulsars, or whatever consumes the limited power reserves while transferring heat from hot objects to colder ones. When the power is gone, the universe should be dead. The stars and galaxies will be extinguished and the pulsars will slow their spinning. On this view it seems that the universe began at some distant but measurable point in time, and it is merely our luck that the show still goes on. I am momentarily reminded of my Rice colleague Larry McMurtry and his novel about a small Texas town, but I cannot believe there will really ever be a *Last Picture Show* for the universe. Am I just blind, or guilty of fuzzy thinking, or is there really some inexplicable problem here?

Philosophically I have preferred the "steady-state universe." In it matter grows old and dies, as we know it must, but, by an act of creation, new matter is created to replace the emptiness caused by the expansion of the universe. This would allow the average density of the universe to be always the same, even as it expands. Any large volume within the universe would then contain galaxies both old and young, just as the human population does. Unfortunately, the state of the universe is a matter of science rather than one of philosophy. The facts make it very difficult for the steady-state idea. While Fred Hoyle was visiting me at Rice University during November 1973, I asked him how he now felt about it. While acknowledging the idea as one of his most satisfying, Hoyle also acknowledged that it was too simple a hope—that the truth always turns out stranger than our best preconceptions of it. Most of the evidence against the steady-state may not actually stand up to intense criticism, but Hoyle said: "I had a very bad feeling about the 3° K background radiation right from the beginning."

The 3° K, or more precisely 2.7° K, background radiation was discovered about a decade ago by research teams at Bell Telephone Laboratories and Princeton University, and it has dramatically influenced everyone's view of cosmology. Like the darkness of the night sky, this radiation tells something profound about the universe, although its interpretation is not unambiguous. The observations, carried out with antennae capable of measuring the flow of radiant energy through the universe at wavelengths between 1 millimeter and 1 meter, are just what one would have expected if the Solar System had been enclosed within a large box whose walls were maintained at a temperature of 2.7° K (that is, 2.7° above absolute zero.) That's quite cold, but its coldness is not the point. The point is that the energy at different wavelengths follows the statistical pattern of maximum entropy. It involves a problem solved by Planck and Einstein near the beginning of this century; namely, if one maintains the total energy of radiation within the box at some fixed value by holding the walls at a fixed temperature, what is the most probable (that is, *most random*) distribution of energy with wavelength? To tackle this problem, they assumed that electromagnetic radiation has corpuscular features, such that each photon (corpuscle) of radiation has a single indivisible quantum of energy, the quantum being proportional to the frequency of the photon. The number of photons in the box was not fixed. They could be created or destroyed at will; but the sum of the energies of all the quanta was required to equal the total fixed energy of radiation in the box. That energy was to be determined by the temperature. They then used laws of chance to find the numbers of photons at each frequency which would give the same distribution (like "two heads and two tails") in the largest number of indistinguishable ways. It was like flipping coins with new rules for radiation invented by Einstein. That may seem a funny way to do theoretical physics, but so powerful are the statistical arguments that they produced a theory of thermal radiation that has been demonstrated correct by countless careful measurements in laboratories. It is one of the landmarks of physics.

Two questions immediately suggest themselves for a universe filled with such radiation. The first concerns how the radiation became distributed in the most random manner when all the known objects within the universe emit radiation predominantly at much shorter wavelengths. The second is the question of running the film backwards. To the first it can be stated that no detectable sources of this radiation can be found. It seems to come from everywhere in the universe, having the same intensity in all directions as maximally random radiation should. Although the radiation is cold by human standards, its total energy is actually large since it fills all of space, requiring its source to be very powerful. Hoyle had hoped that this power could be continuously provided by the fusion of hydrogen into helium because it did seem a curious accident that the energy density of this radiation equals that implicit in the observed average density of helium in the universe. The difficulty is that the energy from thermonuclear fusion is emitted as starlight, whose wavelength distribution is very far from the random 2.7° K distribution. As an explanation it lacked observable means for randomizing the starlight—a tough requirement in an almost empty universe!

In facing the second question, the expansion of the universe comes into play. Because the expansion slowly but surely *lengthens* each wavelength, the film in reverse would show each wavelength shortening. It also shows the number of photons per unit volume to increase because the universe shrinks while the number of photons is preserved. Here a miracle of statistical physics occurs. With each wavelength shortened by the same factor and with the density of photons increased, the distribution remains random. Only the radiation temperature is higher. The radiant energy density increases by a very much bigger factor than the temperature in just the way it must to follow the statistical theory of radiation. This is the single clue that the world of contemporary science seizes upon. It identifies this energy as being heat rather than some other lower-entropy form of energy. It means that the universe would have been continuously hotter as one contemplates earlier and earlier epochs. The situation ultimately changes in this imagined sequence when the contracted universe is so dense that the radiation could no longer travel freely within it. At earlier times the photons are created and absorbed, and the intense radiation creates particles and antiparticles in a still random universe so hot that nothing noteworthy occurs within it other than the chaos of heat. On still earlier frames of the film one sees ordinary nuclei squeezed out of existence in a transformation to heavier elementary particles. It is called the Big Bang because this dramatic film pictures the universe exploding so suddenly from this dense hot state to its present thinned-out proportions. In this view the 2.7° K radiation is just the chilled residual of that primeval fireball. It is the simplest explanation of the existence of the thermal radiation and the strongest evidence for the Big Bang.

The existence of this relict radiation could allow an interesting observation—that of the earth's motion with respect to the bulk of the universe. Because the thermal background is believed to arise from very distant and ancient hot matter of the universe, it should be arriving in equal intensities from all directions. One says that the radiation should be *isotropic*. But if the earth moves with high speed through this photon gas the universe should look hotter in the direction of the earth's motion because the Doppler shift of frequency causes rays moving opposite to the earth to have higher apparent frequency than those moving in the same direction as the earth. The concept is interesting because it seems a throwback to ideas of absolute motion—to the idea of being able to say that the earth moves in a specific way with respect to some cosmic background. Therefore, this observation, which is possible only by virtue of the earth's average motion with respect to the average distant matter, gives the nearest thing to an absolute motion of our solar system. Einstein, with his fascination with Mach's principle, would have been greatly interested. The observation has not proved easy, however; at the present time it appears possible that we move about 200 to 300 km/sec through these photons. Though it is a high speed by human standards, that would not be large by cosmic standards. Indeed, it is surprisingly small, being comparable to the speed of the Solar System's revolution in orbit about the center of our Milky Way Galaxy. The low limit on the speed reflects the smooth isotropy of the radiation temperature, a fact that certainly strengthens its interpretation as the residual of that early fiery furnace.

It's an interesting film—but still only a hypothetical one run backwards from today's snapshot. It does not in itself account for the amazing variety in the night sky. It is particularly difficult to understand how the galaxies and clusters of galaxies formed in this picture of an expanding hot gas; but that is a fundamental problem in any cosmological picture. However, the apparent age of our galaxy does fit into this picture. Can that be a coincidence?

When I was a research student at Caltech, the background radiation had not yet been discovered, although it had been anticipated long before by the late great George Gamow and then forgotten. Nonetheless, we all realized that with the galaxies moving apart and located where they are now, they would have been much closer together in the past. A simple calculation showed that the galaxies would have been essentially touching each other about 12 billion years ago (although at the time we thought it would have been about 7 billion years ago due to the less accurate measurement then available of the rate of the expansion). During these years I began research on the age of the elements within our own galaxy and found the oldest of them to be about 12 billion years old as well—give or take a couple of billion years. It therefore seems, plausibly enough, that the oldest elements within the galaxies were synthesized at about the same time that the galaxies could have come into individual existence. To complete the coincidences, it was becoming apparent at about the same time that the oldest clusters of stars found within the Galaxy, the so-called *globular clusters,* also have an age near 12 billion years. The fact that these three ages agree within this picture seems to lend it credence. When the film is now rolled back only 1 billion years further, the time of that great hot fireball that is now the 2.7° K background seems to be reached. I can only shake my head with wonder that the fantastic complexity that we see all around us could have originated so long ago in such chaotically random beginnings. I can only shake my head in disbelief at a thermal death for the future universe: no more starlight from cold massive dead balls whose fuel is gone; no more high temperature energy for organic life—only forever frozen DNA on dead planets; only cold and darkness and mindless expansion into the black emptiness.

Maybe the film will run backwards. Perhaps the expansion will ultimately come to a halt and the universe will slowly begin to contract. The cold waves will become hotter and the universe will warm up—but with one important difference. There will still be no twinkling stars and no life. The cold death would become only a warm death, and ultimately a hot death. The second law of thermodynamics—or more precisely, "the odds"—defy the reemergence of the low-entropy configurations that now surround us. The situation can be likened to the rapid combustion of gasoline by a spark in a huge insulated vertical piston on earth. The chemical fuel makes heat which rapidly presses the piston to greater height until it expands no more, supported by the hot steam and carbon dioxide within. Even if the cylinder is recompressed, the steam and carbon dioxide will not revert to gasoline and oxygen. The film just will not run backwards. Fortunately, astronomers search for the facts and are not content with philosophy. Huge telescopes around the world search for signs that the expansion slows down. The evidence is inconclusive, but it is out there, waiting like a harvest to be gathered.

I, like you, have the feeling that time is passing. Spring turns to summer, and the pendulum of my old clock ticks in its petty pace from day to day. But what is this thing *time?* As crucial as it is to science, it is regarded in widely different ways by scientists. In particular, what does it mean that time runs in only one direction, unlike space which runs in all directions. Is this a fact of nature or only an idiosyncrasy of the human brain? There are strange possibilities—important possibilities whose implications are unknown. If the direction of time were reversed in the universe, so that it contracts, then *perhaps* the direction of time would also reverse in the Second Law. Perhaps disorder would by itself assume order, and heat would flow from cold to hot while you and I grow younger. Then the film would actually reverse. There's nothing in our laws that prevent it—except the odds. Here is a question for the future: Is the expansion of the universe intimately connected to the observed tendency toward disorder, and vice versa? It is the hardest kind of question, for its answer cannot be obtained by doing the experiment.

Thermodynamics is not the only place where laws are time symmetric in detail while the facts they describe are not. Quantum mechanics has imposed similar features onto the facts of atomic collisions. Suppose a force between two particles is the same regardless of the direction of time—comet Kohoutek scatters past the sun, for example, and exits. If, at its final position, its velocity were exactly reversed, it would exactly retrace the path that even now glows disappointingly dimly above me. Yet when an electron scatters from an atom, the same is not true. If I first find where it is and then reverse its motion, it has only a certain chance of returning to its original point. In this way the process of observation and measurement imposes a time irreversability on *the facts*, although *the equations* describing the motion retain their time symmetry. There is a curious difference between the reversability of the *equations of motion* and the reversability of *the facts*. One of the first examples I ever learned in the university has to do with the certainty with which one can know the location of an electron. As an electron moves undisturbed through space, its future position becomes more uncertain than its position I now measure. The more accurately I measure its position now, the *less* accurately I know its future position. This is the *uncertainty principle* of Heisenberg, and there seems no escape from it. Countless brilliant human work-years of thought and experimental effort have sought without success to circumvent it. Einstein tried very hard. This fact of nature which distinguishes future from present has an interesting relationship to the increase of disorder.

Consider an air mixture whose molecules are half oxygen and half nitrogen, and arrange (perhaps with a removable partition) that the oxygen molecules are originally in one half of the box and the nitrogen in the other half. The dividing partition is then removed. This initial condition is not at all random, for the molecules of different types are in different halves of the container. The Second Law states that in the future the distribution will be more random, and, sure enough, we eventually find that the molecules have diffused and scattered one from another in such a way that both kinds of molecules share equally the entire volume of the container. The film in reverse would be so improbable as to be funny. Well-mixed molecules would be seen scattering countless of billions of times from each other until, lo and behold, the different types appear exclusively in different halves of the container. Each collision obeyed the laws of physics, both in forward and reverse gears, and yet the reversed film is reasonably regarded as physically impossible. Now the usual explanation of the futility of trying to separate the molecules is that one cannot arrange the initial velocities and positions *accurately* enough that the molecules will separate—that it is possible, but just hopelessly difficult. But the uncertainty principle seems to render this statement even stronger—*it is impossible.* One cannot allow the gas to mix, then measure as best one can in principle the positions and velocities of the molecules, then reverse the motions of the molecules, and then expect the gas to return to its initial configuration. The universe evolves irreversibly.

The really curious thing in all of this is that the configuration would be reversible if we did not look at it—if we did not measure the positions and velocities of the molecules. The universe becomes unidirectional in part because we observe it! This seems to me very strange, even after 20 years of worrying about it. It seems to involve the mind of man in the unidirectional evolution of the universe. But the mind of man *is* part of the universe. We are not apart from that which we observe—or *are* we?

> *What is a man,*
> *If his chief good and market of his time*
> *Be but to sleep and feed? a beast, no more.*
> *Sure, he that made us with such large discourse,*
> *Looking before and after, gave us not*
> *That capability and god-like reason*
> *To fust in us unused . . .* Shakespeare

It is probably a simplifying characteristic of the brain that causes us to also look for symmetry. As a schoolboy I worried a lot about the problems of symmetry in baseball. I can't say why except that baseball was of obvious importance, and it therefore seemed worthwhile to worry about its deeper aspects. This I did with a dedication and enthusiasm that no doubt seemed strange to many about me. I worried about the short left-field fence at Fenway Park and the equally cheap shot down the right-field line at the Polo Grounds —places I had never seen but that were very real in my mind without any aid from television. What one wants, I reasoned, are symmetrical playing fields like Dodger Stadium, built in Los Angeles while I was a graduate student at Caltech, or the Astrodome, built in Houston while I was an assistant professor at Rice. I watched the construction of these modern wonder stadiums with remembrances of this childhood problem, and it is a quiet disappointment of my adult life that baseball has never seemed as exciting and wonderful in them as it had in those magical old parks. I would suppose that's just age, except for the equal lack of enthusiasm I see in most kids today. Perhaps it's a different world—one that has made a phase transition with respect to baseball! Nonetheless, I had once found it to be of great importance that the left and right probabilities be equal. It is, after all, one of the joys of baseball that it is a statistical game—a game where the great hitter strikes out and the best team does not always win. In baseball, truth is statistical, but surely it should be statistically equal for left and right.

I soon realized that baseball cannot be symmetric. There are more right-handed pitchers than southpaws, for one thing. That's just a fact. The left-hand batter can run to first base more quickly than the right hand batter, whereas a good right hand hitter necessarily puts more pressure on the shortstop than the left-hand hitter ever can on the second baseman (Ted Williams excepted!) The fundamental asymmetry of baseball lies in the fact that the bases are run counterclockwise. Only if the rules were changed could baseball be symmetric. The simplest expectations are often shattered by the facts.

In the very year I entered graduate school Yang and Lee made their Nobel Prize winning suggestion that nature itself has a similar asymmetry. At the time I could not comprehend how nature could possibly distinguish between "right-handed" or "left-handed" events, considering that these seemed artificial distinctions of the human mind. I believed, as I had also been taught, that natural events viewed in a mirror obeyed the same laws. The motion is clearly changed by the reflection, but in such a trivial way that the laws describing the motion should surely remain the same. They do not. A cobalt-60 nucleus spinning counterclockwise (like the base runners) emits its radioactive beta particles into the ground rather than into the air. That may seem innocuous enough, but I have spent hours convincing myself that if I arrange things so that the cobalt 60 when viewed in the mirror appears counterclockwise (like the runners), the beta particles are emitted into the air! This shattered expectation is called the "violation of parity" by physicists. It sounds pretty serious. For human thought, it is.

Since then I have continued to believe, but with caution, that the basic microscopic laws are time symmetric. Surely if we run the film backwards, the laws of physics are the same even though the motion differs. Oh sure, very improbable things will happen with large numbers of particles, but they are only *very improbable* rather than *impossible*. One could get all heads, I suppose. Oh, yes, there is some problem with quantum mechanics if we measure the final state before reversing the film, but in principle we can run quantum mechanics backward if we don't look first at the forward film. Still, it has seemed possible that the basic microscopic laws do not notice the direction of time. However, it now seems probable that they do. Experiments on the weak decays of mesons indicate a weak violation of time-reversal invariance. If further experiments substantiate this fact unambiguously, we will at last have to admit that even at the submicroscopic level nature knows irreversibly which way time is passing. Some weak decays proceed only one way, order passes to disorder, and the universe expands. Meanwhile the elementary particle physicist is being pushed toward a final symmetry: If we view nature in a mirror, *and* replace all particles by their antiparticles, *and* reverse the direction of time, the laws of physics are the same (we hope).

What has all of this to do with cosmology? I actually am not sure, but it seems reasonable that those conceptions of nature that reflect in part the human brain also reflect that brain's image of the universe.

CHAPTER XV
THE EYES OF TEXAS

In November 1972 my wife Annette and I were on our way to the Big Bend National Park. We had already driven almost 600 miles from Houston and were on the last leg south from Marathon when our eyes saw three vultures circling high above the road ahead. From their dizzying height their piercing eyes were focused on something on that road. When we got there we found a magnificent rattlesnake killed by a car. The vultures continued to circle—watching. We decided to watch too, and drove our car a small distance away and waited. The vultures are accustomed to cars now. They have adapted somewhat to this mechanical friend, and their keen eyes scan the highways of West Texas continuously. Suddenly they decided it was safe. Down they circled and, before our amazed view, picked the snake clean. I photographed one of them ripping off a nice chunk.

After passing Persimmon Gap another 40 miles to the south one finds a raw wilderness. I know of nothing quite like it anywhere. From the ascending road toward Panther Junction, the primitive Terlingua flats cause inner chords to tremble. Maybe it's because I read *African Genesis* and can never forget it, or maybe it's because the "Dawn of Man" in *2001* could almost have been filmed at Terlingua, but my eyes stare through that highland desert transfixed. I cannot but think we were all there long ago. My eyes! Like the vulture I would be useless without them—at least in this ancient setting. I augment them with camera and binoculars. I am the equal of the vulture as I pull my 7x35's to my eyes and just stare across that wasteland. Power and knowledge fill me as my eyes probe distant bushes and ravines. A sense of well-being inexplicably fills me at this power. I see, and I comprehend.

Our genetic roots in this experience must be very strong. I have always marveled at the preeminence of visual perception over the other senses. When I'm introduced to people, their names usually go in one ear and out the other. They don't go straight to the right part of my brain. But if I read a person's name on a piece of paper, I feel I will remember it forever. To do mathematics I have to write it down and look at it. If I really want to remember a name I hear, I imagine it written on a piece of paper where I can see it—and then I relax and remember. It goes to the right part of my brain. The examples are legion. In physics courses at Rice University the students (and sometime I) struggle to visualize the invisible: what does the atom look like? It boggles the mind to realize that the atom could not be "seen" even by the keenest of eyes. It's impossible even in principle, although the electron field emission microscope comes close, but still we press on trying to get a visual picture of something with no visual form. Why are we like this? Our instincts seem to know two kinds of knowledge: a superior kind that comes through the eyes and an inferior kind that comes in any other way. Our intellects try to be sophisticated, but our instincts continually betray our association of knowledge with our eyes. The answer must be that our visual sense has a more direct route to the important portions of our brain; and that's not surprising, considering the importance of visual information to the evolution of the species.

Exactly one year later we returned to the Big Bend with Fred Hoyle. I was in part eager to repay him for my own introduction to the Scottish highlands. As we took a shortcut across one of the switchbacks on the Lost Mine Trail, Annette leaped in shock at a huge fuzzy brown tarantula at her feet. The creature looked frightening, and, although its bite is not actually poisonous, we instinctively gave it respectful distance. Fascinated, we watched it from about 1 meter for almost half an hour. What is the meaning of this fear of spiders—even fear of harmless spiders? It is so common in mankind. We see the tarantula and pull back quickly. Almost no one has the same reaction to birds. Is this something we learn, or, as I suspect, does the visual image meet a warning pattern in our brain that was fashioned there long ago in the evolution of the species?

Vulture on the road south from Marathon

Can DNA actually contain that much information? Why do the gazelles and giraffes of Africa graze tranquilly together, but flee at a glimpse of the silent lion? These tarantulas abound in the Big Bend, in equilibrium with their environment. But the days of this individual huge spider that we then were watching may be very few in number, for the blue wasp may soon see him. She paralyzes him with her sting and then plants her eggs within his huge fuzzy body. They literally devour the dying spider.

The cytochrome of the human cells is a chain of 104 amino acids. It plays an essential role in all cells in the metabolism of nutrients to provide energy. The number, sequence, and frequency of the amino acids determines the nature of the proteins, and in this case reveals genetic similarities. The cytochrome of the Rhesus monkey is an identical chain of the same 104 acids. Evidently our kinship is close. The cytochrome of the horse has twelve differences from the human in this chain of 104 acids. Even in such a remote species as the invertabrate moth, there are only twenty-six differences in the chain, plus a modified structure at its end. The similarities are very great and show our evolutionary relationship to even the moth. From the number of differences it is clear that we and the Rhesus diverged from the ancestors of the horse at a much later time than from the ancestors of the moth.

In all the lower species we easily identify the crude instinctive patterns for survival—irrational except by nature's impassionate standards. It is curious that we view ourselves so differently—rational, logical, and omniscient. But might we not appear as amusingly pathetic to another civilization *out there* as the baboon (a very high form of life) does to us? If I have any belief, it is that the only good future for man is to rationally identify himself as part of nature rather than to maintain his belief in a godlike dominion over it.

More to the point is the question of the brain. If we could but learn how the Rhesus monkey thinks, would we not find a cruder version of ourselves? Does he not also build conceptual patterns that aid his interpretation of his environment? To each of us as infants the world was a chaotic blur, and we each had to invent thought patterns to render it meaningful. And first and foremost of our education, probably the most important step we take, is to learn to focus our eyes. Because vision opens such a floodgate of knowledge, it is not surprising that it never loses its preeminence as a form of knowledge acquisition. Geometry and an instinct for mechanics quickly follows. Having two eyes instead of one allows a perception of depth by an elementary—one might almost say *biological*—application of trigonometry. One can only shake his head in wonder at the marvelous swift calculations of the Rhesus brain as he easily catches the thrown peanut—even hanging by his tail! We must continually remind ourselves, if we are to appreciate what we are, that this reaching out to catch the peanut at a point in space-time on its parabolic orbit is *not* the simple instinctive reaction it seems to be—it is a marvelous calculation by a computer trained to recognize the pattern and instantly to compute. The pattern recognition becomes valuable, and the recognition of patterns is aided by concepts—momentum, energy, time, space, redshift, parity, inertial frames, symmetry, antiparticles, love, hate, and me.

This disposition toward conceptual structure is in all of us, but we must each teach ourselves to use it. We often naively suppose that facility at mathematical thought is inherited in detail—that one person is just "smarter" than another due to a gift of birth. Undoubtedly true for the human species *vis-à-vis* other species, I think it is probably not true of the differentiation within our species. As our blank minds begin to handle quantitative concepts, we each face countless choices in writing the programs for our own mental computer. There are several different logical ways to conceptually regard the quotient of 64 by 8, for example, and each of them has a fascinating logical appeal to the person who discovers it. The young mind probably tries experiments in the attempt to find logical patterns capable of swift correct answers. As the number of logical operations and choices becomes large, so too do the degrees of differentiation among human brains. The thought that there is "one way to multiply," and some learn "it" better than others shows an improper appreciation of this great human accomplishment. The difference between Fred Hoyle and my grocer is not so much a difference of inherited ability as it is an indication of the many fortuitous early discoveries that today give Hoyle his astonishing mental patterns.

I cannot but in this regard think of the mathematician G. H. Hardy, who helped shape the Cambridge I came to know. In *A Mathematician's Apology* Hardy discusses brilliantly and clearly the mathematical life. He also describes his undisguised admiration of the Indian mathematical genius, Ramanujan, who knew deep mathematical truths without even knowing how he knew them. Indeed, according to Hardy, Ramanujan hardly understood what a "proof" constituted to a formal mathematician. He just knew the answer. He died a tragic early death at age thirty-three, and Hardy recalls visiting him near his end and remarking that the license plate of his taxi was 1729—hardly an interesting number. "No, Hardy! It is a very interesting number," Ramanujan replied. "It is the smallest number expressible as the sum of two cubes in two different ways." [$1729 = 12^3 + 1^3$ and $1729 = 10^3 + 9^3$.] Who are we to fathom the rare combinations of mental logic that makes facts like these so transparently obvious to a man like Ramanujan? Great mysteries lurk here.

So quixotic is the mathematical imagination of the human brain that it sees truths that it cannot prove, at least not without years of frustrating effort. For example, many fine mathematicians have tried for more than a century to prove a hypothesis of Riemann to be correct. Technically, he conjectured that the complex (i.e., nonreal) numbers for which a specific function, named the "zeta function," is equal to zero are numbers which all lie on a critical vertical line in the complex number plane. That seems esoteric enough, but many practical problems in number theory hinge on its correctness. Although it has been believed correct for more than 100 years, an army of mathematicians has been frustrated in its attempts to prove it to be correct. What exactly is this brain, with its logic and its numbers, that it can torment itself with unproven truths? And what is the relationship, if any, of this kind of mathematical intuition and truth to the intuition of physical truth about the real world? Do our minds shape our world, or does the world shape our minds, or both? In such profound confusion we confront the problem of the meaning of our existence.

With these brains of ours we look outward to the universe for understanding, just as we did from the dim dark caves of our past. First with the eyes, then with enlarged eyes, or telescopes, and now at last we "look" with every type of information that comes our way. What we find remains a mysterious combination of what's out there plus our own minds. It is a great thrill to be a small part of this adventure. To press on with it we will need new eyes—not the eyes of man, but eyes for mankind. About a hundred miles to the north of the Big Bend the new University of Texas Radio Observatory is measuring day and night the structure of distant sources of radio waves with angular accuracies of ten seconds of arc. About 150 blue stellar objects have been confirmed as being QSOs. Not far away the famed telescopes of the McDonald Observatory in the Davis Mountains continue their classical seeing with new accuracy. The newest, the 107-diameter reflector, was supported primarily by NASA for the observation of the planets. At Rice University my colleagues Low and Haymes have flown infrared detectors and gamma-ray detectors respectively to the top of the earth's atmosphere—one in a Lear Jet and one in a giant helium-filled balloon. And at Clear Lake, south of Houston, the "giant step for mankind" was planned and trained, and the Skylab crews conduct new types of observations of the universe around us. We will not quit this thrust while we remain men because we are a part of that great universe, and we must know what it is.

I know that the ideas I have been sharing with you will never leave my life. They will follow me to the end of my days. These questions possess my soul, and if they do not give me peace they nonetheless

The Rice University gamma-ray telescope parachutes back to earth from a balloon suspended atop the earth's atmosphere into the desolate badlands of west Texas.

give me joy. They make me proud to be a man. They provide the fun of stretching my mind to grapple with, and sometimes understand, something new. And most of all, this quest endows life with that undefinable quality that makes it all seem meaningful. Even now, in Spring 1974, my parents' maple tree, which I planted 25 years ago, leafs anew. It will be more intricate than last year, and more exquisite.